ECOLOGY
of an
UNDERWATER ISLAND

ECOLOGY
of an
UNDERWATER ISLAND

Robert W. Schmieder

Cordell Expeditions • Walnut Creek, California

Cover: Randy Schmieder, 1990

From hundreds of underwater photographs, the artist has captured the spirit and visual impression of a typical scene on the bottom at Cordell Bank, where the cover is so dense that not a single spot of bare rock shows. The sponges, crabs, anemones, hydrocoral, hydroids, urchins, snails, worms, and fish shown here are seen as the author saw them on hundreds of dives over a ten-year period.

Library of Congress Catalog Number: 90-082110
ISBN: 0-9626013-0-6

Printed in Hong Kong

CORDELL EXPEDITIONS
4295 Walnut Blvd.
Walnut Creek, CA 94596

To my parents ...

OTTO and RUBY SCHMIEDER

... thank you both for a lifetime of love.

PREFACE

This brief book is really a case study: a summary of what we know about the ecology of a rocky bank 100 km northwest of San Francisco. Cordell Bank is well-described as an *Underwater Island*, by which we mean a limited area within which conditions are favorable for supporting a rich and vigorous biological community, but surrounded by a barrier that causes the community to live in relative isolation.

Until very recently, Cordell Bank was practically unknown even to the scientific community. Partly as a result of a recent series of expeditions and partly as a result of assembling existing information, we now know enough about the Bank to realize that its lushness results from the overlap of several favorable oceanic conditions. Even a brief study of this site would convince one that it represents an extraordinary resource for conservation and research. Recognizing this, the National Oceanic and Atmospheric Administration (NOAA) has established Cordell Bank as a National Marine Sanctuary, which includes not only the Bank itself but also the surrounding waters.

Most of what we know about the environment at Cordell Bank resulted from a series of exploratory expeditions led by the author and colleagues. From those expeditions, we were able to make an initial description of the community, and take the first steps toward understanding what controls it. This book summarizes the physical factors that appear important to the biotic community living on Cordell Bank, and how the community interacts with itself and its surroundings—the *Ecology of an Underwater Island*. To the extent that Cordell Bank is a prototypical hard-rock, shallow-water bank, these observations should have relevance to many other similar sites.

This is a story about relationships and limitations in a specialized oceanic habitat, the shallow subtidal rocky bank. Thus, we expect to see many of these same relationships and limitations on rocky banks off New England, Mexico, South America, in the South Pacific, and in the Indian Ocean. Not everything everywhere will be the same, of course, but what we have learned of Cordell Bank will help us understand what we see at these other places.

CONTENTS

ACKNOWLEDGMENTS

It is with considerable pleasure that I extend to each of the many dedicated people who have contributed to the material summarized here my sincere and heartfelt appreciation.

The Cordell Bank project was conceived by the author with John Hall. The first expedition to the Bank was carried out by the author with Larry Pfoutz, Dannie Baxter, Don Griffin, and Steve Lawler. Many additional divers deserve the credit for the subsequent underwater exploration: Kathryn Amenta, Gary Borton, Phil Byrne, Dave Cassotta, Gordan Chan, Susan Dinsmore, Don Dvorak, Sue Estey, Angelo Festa, Paul Hara, Bob Hollis, Dale Howe, Bill Kruse, George Lang, Doug Laughlin, Jennifer Linton, Jack Marshall, Steve McCormick, Lee McEachern, Steve Metzger, Rick Moore, Rob Morris, Doug Niessen, Ron Owen, Barry Parker, John Payne, Chuck Rawlinson, John Santilena, Tom Santilena, Carl Schmitt, Jerry Seawell, Harry Sherman, Skip Smiley, Lew Stark, George Styer, Lori Talbot, Erik Taylor, Lee Tepley, Ben Tetzner, Dave Walls, John Walton, Steve Wedi, and Steve Williamson.

Among these I wish to especially note the extraordinary contributions and long term continuing work on the project by Don Dvorak, Sue Estey, Bill Kruse, Tom Santilena, and Harry Sherman. Their continuous unrestricted dedication, resourcefulness, and enthusiasm were crucial many times in many places. And among these, the person most responsible for the success of the project was Bill Kruse. Besides his diving, photography, logistical, and strategic contributions, he is responsible for creating the spectacular computer graphics depicting the topography of Cordell Bank.

Special thanks are extended to the skippers of the various vessels used on the expeditions: Mike Craine, Clay Elmore, Wilson Landrum, Bob Landrum, Breck Greene, Bert Mosca, and Chuck DiGiulio. I am particularly grateful to the dozens of volunteers who helped prepare the author's vessel, the *Cordell Explorer*, for sea duty, especially Gary Borton, Sue Estey, John Esterl, Keith Flood, Dale Howe, Tim Howe, Bill and Gail Kruse, Brian and Tommye McGuire, Randy Schmieder, Russell Schmieder, Elaine Senf, Harry Sherman, and Melanie Stright.

Specimens from Cordell Bank were identified by Paul Silva, Richard Moe, Terrie Klinger, John LeClaire, and Kathy Ann Miller (algae); Al Mahood, Sam VanLandingham, James Fidiam, and Margaret Hanna (diatoms); Welton Lee, Barbara Bowman, and Martha Spence (sponges); J. Wyatt Durham, Steve Cairns, and Nan Chadwick (hard corals); Lyn Raible, Joan Steinberg, and Mary McGann (microfauna); Robert Higgins and Marie Wallace (meiofauna); Barbara Weitbrecht (polychaetes); Dalene Drake (amphipods); William Newman (barnacles); Barry Roth, James McLean, Frank Bernard, Thomas Waller, David Behrens, and Barry Putman (mollusks);

Dustin Chivers, Dan Gotshall, Robert Lea, John DeMartini, and Robert Given (fish and invertebrates); Ann Muscat and Tim Coffer (brittle stars); Steve Cooper, Michael Ellis, Bill Keener, Hal Markowitz, Izzy Szczepaniak, and Marc Webber (birds and mammals); and Dave McCulloch, Don Ross, and Jim Mattinson (rocks).

Facilities and/or services were provided gratis by Marine World/ Africa, USA, TDC Underwater Systems, Anchor Shack, Monterey Diver Center, Texas Instruments, Continental Color Labs, San Francisco State University, Bodega Marine Lab, Fairchild Camera and Instrument, Lawrence Livermore Laboratory, and the California Academy of Sciences.

Financial support for the Expeditions was provided by the National Geographic Society, the Conservation and Research Foundation, the Explorers Club, the San Francisco Foundation, and the Sanctuary Programs Division of the National Oceanic and Atmospheric Administration (NOAA), as well as many individuals. NOAA staff who were helpful include Chris Andreasen, Nancy Foster, Bruce Hillard, Millington Lockwood, Ralph Lopez, Richard Perry, Don Pryor, and Tom Richards.

Advice, encouragement, and liason at various times and places was generously provided by Gordon Chan, Dustin Chivers, Sylvia Earle, Gary Fellers, Bob Given, Michael Herz, Dan Gotshall, Hal Markowitz, Dave McCulloch, and Phil Trupp.

Staff at the Whale Center, Oakland, especially Maxine McClosky and Mark Palmer, provided the vital link between the public and the U. S. Congress to help include the proper protection for the Cordell Bank National Marine Sanctuary. In the final stages of the designation, Congresswoman Barbara Boxer, Congressman Doug Bosco, and Senator Pete Wilson provided crucial influence to complete the protective legislation. President George Bush signed the final Act of Congress providing complete legislative protection for Cordell Bank.

Exceptional appreciation is extended to Paul Silva and his student and co-worker Richard Moe for extraordinary help and constructive criticism on this manuscript. Paul, especially, provided uninterrupted support from the first day of the 10-year project to the last.

Others who helped with preparation and technical editing of the manuscript include Alice' Davis, Martha Franks, Kathy Kramer, Ken LeJoie, Mary McGann, Jim McLean, Charles Powell, Randy Schmieder, Jane Staley, Lori Talbot, and Mary Wicksten. Illustrations and photographs were generously supplied by a variety of dedicated people; they are listed just after the bibliography. In particular, the author's younger son, Randy, created many of the complex graphics and provided considerable editorial help during the preparation of the book.

To all these, and any I may have missed, my compliments and sincere thanks; please keep in touch.

RWS
Walnut Creek
1 September 1991

PROLOGUE

... to keep you from gettin' more nervous than you already are, I'd be willing to wager you that before this night is over, that-there chunk of lead is going to fetch up down below in thirty fathom of water. Rocky bottom with live barnacles. And that'll be the Cordell Bank: God's own gift to Frisco-bound shipmasters in this kind of weather!

—*Voyage*, by Sterling Hayden, 1976

BY KATHLEEN MACLAY
Times Staff Writer

WALNUT CREEK — A local physicist turned avid amateur will lead an expedition to explore a shoal at least 100 feet under water west of Point Reyes that has been left virtually uncharted since its discovery more than 100 years ago.

Dr. Robert Schmieder, an atomic physicist at Sandia Laboratories in Livermore, along with fellow divers, wants to be among the handful of adventurers who have dipped to the depths of the Cordell Bank.

On the trip planned for Sept. 6, 8 or 11 — depending on the weather — they hope to find specimens of invertebrate sea life, algae and maybe even relics left by coastal Indians who probably used to visit the once submerged Cordell Bank area.

Schmieder doesn't expect to find another Atlantis, but on the other hand, he doesn't know what the crew might turn up.

Information about the bank located about 20 miles west of Point Reyes is scarce. In some cases it's top secret. It is known that during the last Ice Age, the ocean level was several hundred feet lower than it is now. Cordell Bank was an island at that time.

"It would have been much like Mt. Diablo," Schmieder said, adding that some portions of it seem, according to available data, to jut upward in towering peaks.

"It would have been much like Mt. Diablo," Schmieder said, adding that some points on the bank appear to jut upward like mountain peaks. He reasoned that the coastal Indians could have traveled easily to the island, which likely supported a large amount of bird life and thus provided the people with eggs and fowl for eating.

The first official recorded report of Cordell Bank was in 1853. A seafarer named George Davidson noticed that his boat was veering off course, by a force he couldn't identify. Later it was discovered that the area, because of its location on the edge of the Continental Shelf and the San Francisco Bay, is subject to a reverse offshore current.

By 1869, Pacific traffic in the region warranted an investigation of the shoal by the U.S Coast Guard.

Edward Cordell, well-known for his oceanographic charting in the Pacific and Atlantic, was assigned to the task of finding and mapping the shoal. Later, it was named in his honor.

Another brief survey was done of the area in 1872 and in the 1920s, general surveying with automatic depth sounders turned up fairly accurate contour mapping of the bank.

But Schmieder said many of the details still

of marine mammal and bird experts to explore and chart the bank.

"The sensation you have is one of floating above the very highest peak of a mountain," Schmieder said. He compares it to being "a bird floating above a high building ... It's a giddy feeling."

The mountain peaks are saturated with brilliantly colored corals, anemones, sponges and hydroids to a depth of about 160 feet below sea level. Divers from Cordell Bank Explorations, a non-profit organization based in Schmieder's Walnut Creek home, have identified more than

mean is that we'll have to get a permit to go diving."

The Cordell Bank could provide Schmieder with enough thrills to last a lifetime. He estimates that his team has explored less than 1 percent of the bank, and every expedition brings some new discovery, however small.

The teams dive during the summer and fall, but Schmieder works on the bank year-round, keeping in touch with experts on marine life and geology, including cataloging and sending rock and species samples for identification or classifica-

ROBERT W. SCHMIEDER shows barnacles and sponge-encrusted coral in ... Cordell Bank, in bottles behind him

tion, and writing papers about the work.

Schmieder, 45, has degrees in physics from the California Institute of Technology and Columbia University, and did postdoctoral work at UC-Berkeley. He relaxes by playing piano, sometimes composing his own music, but his biggest thrill comes from discoveries of the kind he has made on the Cordell Bank.

He sees the process of exploration as common to every field of science, and it is that process, rather than a particular field of study, that fascinates him.

"The thrill of discovery is the bottom line," he said, whether in physics or marine biology. "You sit in the laboratory and the needle goes to a certain place, and you know you've measured something that's never been measured before.

"If you're lucky, you get one small thrill a year. If you're good, you get one big thrill a lifetime."

California hydrocoral clusters on a Cordell Bank granite ridge.

glacier me... dell Bank, land at a... California. Cordell... and 3 miles... sea level, Walnut Cre... five year p... month. For the... spot is mo...

The many riches of

By Spencer Sias
Of the I-J staff

Look at an offshore topographical map and Cordell Bank will pop out at you.

At least that's what happened when Robert Schmieder looked at the marks showing an undersea mountain or ridge 20 miles off Point Reyes more than three years ago.

"It just leaps out at you," Schmieder, an amateur diver and professional physicist, said.

It gripped ... urge to explo... After all, ... about Cordell ... it pushed up ... ocean floor ... surface. Not... ners needed... About the ... had created ... along was a ... son, director ... vey, to nam... discovered ... mapped it r... Davidson ... brand on the ... others named ... who had ex... thoroughly in ... the map. Still, "it w... essentially s... When Schm... formed the C...

Scientists to study 'lost island'

Cordell Bank off Point Reyes

By Fred Garretson

Twenty miles west of Point Reyes, at a spot where the ocean currents sometimes flow backwards, there is a mysterious submerged island known as Cordell Bank.

If weather is favorable, scientists in scuba-diving gear will plunge into the Pacific Ocean in early September to explore the lost island, whose peak sits 120 feet below the sea surface.

They plan to study the geology, gather samples of algae and ocean plants, measure the growth rate of sea-bottom organisms and, perhaps, find traces of the early American Indians who may have lived on the island before it was submerged.

The expedition will be led by Dr. Robert W. Schmieder of Walnut Creek, an atomic physicist at Sandia Laboratories in Livermore.

Schmieder says the water is clear and "there is lots of light" on the bottom. The hard granite rock peak of the lost island sits in submerged twilight. The sunlight feeds a lush underwater jungle of plants and fish on the slopes below. This heavy biological activity is visible on the surface, coloring the water and attracting unusual numbers of seals, birds and fishing boats to this spot in the open sea.

Cordell Bank sits on the very edge of the continental shelf. Nautical charts show that on the west side of the submerged island there is a steep slope — almost like an underwater cliff — where the seabed drops to 10,000 feet deep in the space of a few miles.

But on the east there is shallow water with a maximum depth of 360 feet. At the end of the last ice age, about 10,000 years ago, when sea level was 300 to 400 feet below its present surface, Cordell Bank probably was a large island about nine miles long and four miles wide, separated from the mainland by a narrow strait of shallow water. For a time, it may even have been possible to walk on dry land from Point Reyes to Cordell Bank.

"Since North America was inhabited at that time, it is likely that it was visited perhaps as a source of food," Dr. Schmieder says.

It took a long time for the melting glaciers to raise the sea level and cover the island with water. Not long before ancient Egyptian civilization was getting started, California Indians would have seen Cordell Bank as an island more than two miles long and two miles wide.

Scientists acknowledge they have only a remote chance of finding the remains of early man during the few hours of scuba diving scheduled for this expedition. But they expect to learn a great deal about the special aquatic environment around the submerged island.

"Fish, including sharks, are abundant," Schmieder says. "Tomales Bay, a breeding ground for Great White Sharks, lies 25 miles to the northeast. No specimens of algae, and only a few invertebrates, from there are known."

He says the first recorded observation of the submerged bank was made in 1853 by George Davidson, "who found himself pushed off course by a mysterious unknown force," identified years later as a reverse offshore current. Other ship captains reported the same "unknown force."

Edward Cordell, an assistant superintendent of the U.S. Coast Survey, finally located and mapped the

Continued on page 4

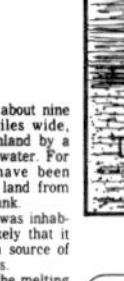

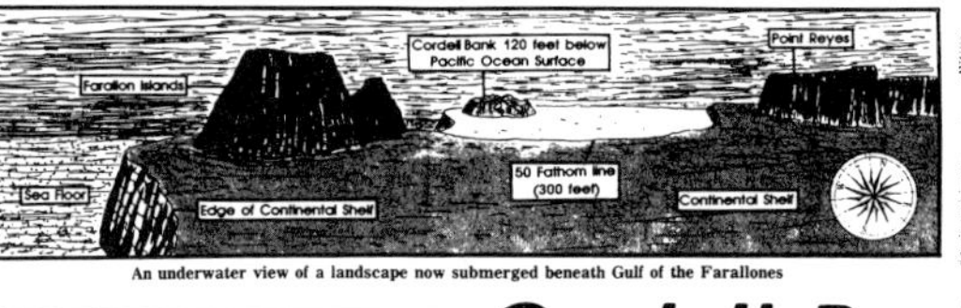

An underwater view of a landscape now submerged beneath Gulf of the Farallones

An amazing island underneath the sea

By Fred Garretson
Staff Writer

Bodega Bay—The Walnut Creek scientist was astonished when he dived to the peak of Cordell Bank, a submerged island in the Pacific Ocean 20 miles west of Point Reyes.

"It is a unique place with an astonishing, unbelievable amount of light," says Robert W. Schmieder, who as leader of a scuba diving expedition last month was the first person ever to dive to the peak.

In addition to many scientific, oceanographic and archeological puzzles, Cordell Bank has an aura of spy-versus-spy mystery.

Russian trawlers believed to be electronic spy ships and international oil explorers have both prowled around the area.

The strategically-located submerged island—ideal, for instance, for hiding a submarine—was the target of a still-secret U.S. Navy expedition in 1969.

...ll Bank

Group, scientists and marine biologists were dubious.

"There were a lot of dire warnings about it," Schmieder said, recalling that one local scientist said, "I wouldn't go there. It's deep and dark and cold and there are a lot of sharks."

"All of those are true," Schmieder acknowledged.

But the urge to explore those waters was greater than any fears stirred up in them.

"We were all amateur divers — sport type divers, all people who are either involved in science professionally or in some technical area," Schmieder said.

"Diving was simply a hobby for us, but eventually you have dived in all of the areas that are financially accessible.

"We just became motivated to do

See Bank, page 6

Cordell Bank divers find s...

By George Nevin

The flat Pacific stretches away, calm and waveless, as the five divers pull on their wetsuits and prepare to slip beneath the surface.

It's a warm, windless October Friday, the perfect day for the five accomplished divers and amateur scientists to visit a place where scuba divers never have gone before.

Their target: a submerged island 20 miles due west of the tip of Marin's Point Reyes Peninsula, a dramatic seamount called Cordell Bank.

What shadowy world lies 150 feet beneath their boat? By all accounts, the seamount is a gloomy jumble of granite peaks and valleys, a dim anachronism looming out of the ocean floor. Sunlight, filtered through 25 fathoms of seawater, varies from dim to non-existent.

Walnut Creek atomic physicist Robert W. Schmieder and Larry Pfoutz ponder these possibilities as they get set to enter the turquoise-green water.

The three other divers — Bob Justice of San Anselmo, Dannie Baxter of San Rafael and Don Griffin of Mill Valley, director of research for the California Marine Mammal Center at Fort Cronkhite — wait quietly for their turn. They will form the second contingent of divers, if there is one.

Just as the divers spash overboard from a little rubber boat, a fog bank materializes. There is momentary worry. Will the expedition, planned for more than a year, postponed 16 times and actually scrubbed twice because of bad weather, have to be aborted again?

But the fog clears, and Schmieder and Pfoutz, a former Navy diver, swim down toward the twilight world of Cordell Bank. It is a journey that will yield treasures to thrill the algaeologist — the scientist who studies algae — and leave the divers with a taste for more.

As the two men descend, they think of all the stories they've heard of Cordell's dark and dreary world.

Let Schmieder take up the story:

"In California waters at 100 feet, normally things are dark — you need a dive light. Beyond that, only a few types of organisms can exist.

"But at Cordell, the water is much clearer, so there's more light. The California Current continually washes the bank, sweeps it, so debris and excreta from organisms don't exist. It's visually different from any other diving spot in California."

For 15 minutes, while their companions overhead sweat out the fog bank that formed and then evaporated, Schmieder and Pfoutz paddle through a submarine world startling in its beauty and compactness.

The rocks are covered with knob-like clumps of tiny anemones; thrusting out of the abundant life are huge white disc-shaped sponges that look like half-buried wagon wheels.

"There was no debris; everything was absolutely clean," Schmieder said. "The organisms were clumped into groups, not distributed chaotically. Usually, you see chaos underwater, but Cordell Bank seemed as though it had been landscaped. Colonies of life forms were stuck together, but they weren't excessively crowded; they seemed comfortable."

Despite what Schmieder called "dire predictions that Cordell was a wasteland because it was deep and dark," he said all the plant and animal life was "healthy and robust, astonishingly so."

Their chosen diving spot was no less astonishing than the life it supported. They had aimed for, and hit, a pinnacle of quartz-diorite thrusting up from Cordell Bank — a 12-mile-long mass of what once was a chunk of the Sierra Nevada.

On one side, the pinnacle sloped down toward a shallow underwater plain that led off toward Point Reyes. On the other, the rock spur fell off in a 200-foot vertical wall, then dropped almost 10,000 more feet in the space of a few miles.

The history of Cordell Bank is a violent one. Torn from the rest of the mounta... years ago ... immense pla... border, the ... rocks of Co... way 350 mile...

Over succ... ciers had fo... ing the ocean... was locked u... and rise whe...

This last c... years ago, ... ice age, caus... by about 400... tainly expos... and may ha... attached to ... west from ... promontory ... insula.

Other sci... shown that C... at the time ... although the... that Native ... Point Reye...

By Bill Kruse

...kes samples on Cordell Bank off Point Reyes — scientists reap benefits of expedition

Visit to a long-lost island

By FRED GARRETSON
Staff Writer

Twenty miles west of Point Reyes, at a spot where the ocean currents sometimes flow backwards, there is a mysterious submerged island known as Cordell Bank.

If weather is favorable, scientists in scuba-diving gear will plunge into the Pacific Ocean tomorrow morning to explore the lost island, whose peak sits 120 feet below the sea surface.

They plan to study the geology, gather samples of algae and ocean plants, measure the growth rate of sea-bottom organisms and, perhaps, find traces of the early American Indians who may have lived on the island before it was submerged.

The expedition will be led by Dr. Robert W. Schmieder of Walnut Creek, an atomic physicist at Sandia Laboratories in Livermore.

Schmieder says the water is clear and "there is lots of light" on the bottom. The hard granite rock peak of the lost island sits in submerged twilight. The sunlight feeds a lush underwater jungle of plants and fish on the slopes below. This heavy biological activity is visible on the surface, coloring the water and attracting unusual numbers of seals, birds and fishing boats to this spot in the open sea.

Cordell Bank sits on the very edge

have been possible to walk on dry land from Point Reyes to Cordell Bank.

"Since North America was inhabited at time, it is likely that it was visited, perhaps as a source of food," Dr. Schmieder says.

It took a long time for the melting glaciers to raise the sea level and cover the island with water. Not long before ancient Egyptian civilization was getting started, California Indians would have seen Cordell Bank as an island more than two miles long and two miles wide.

Scientists acknowledge they have only a remote chance of finding the remains of early man during the few hours of scuba diving scheduled for this expedition. But they expect to learn a great deal about the special aquatic environment around the submerged island.

"Fish, including sharks, are abundant," Schmieder says. "Tomales Bay, a breeding ground for Great White Sharks, lies 25 miles to the northeast. No specimens of algae, and only a few invertebrates, from there are known."

He says the first recorded observation of the submerged bank was made in 1853 by George Davidson, "who found himself pushed off course by a mysterious unknown force," identified years later as a reverse offshore current. Other ship captains reported the

Robert Schmieder of Walnut Creek leads Cordell expedition

portant navigation point for captains caught in off-shore fogs. They would drop weighted lines, detect the shallow water "seamount," and know their position was 38 degree north latitude by 123 degrees 25 minutes west longitude.

Cordell Bank may be important in the era of undersea weapons.

It is a natural barrier blocking sonar devices used to detect the location of vessels and provides a hiding place for submarines. The hard-rock shoal could be used as a platform to hide submerged missiles, underwater detection gear, spy devices and secret radio transmitters designed to float to the surface at pre-arranged times.

Ten years ago a U.S. Navy expedition made a number of dives at Cordell Bank as mysterious as the bank itself.

The Navy issued no reports, took no biological samples, and still classifies the purpose of the expedition as secret.

The Soviets also have a keen interest in Cordell Bank. Coast Guard reports show fleets of Soviet trawlers have stopped at the site several times, sometimes for days at a time. The trawlers, which often bristle with electronics gear, are regarded by U.S. experts as intelligence gathering vessels.

Cordell Bank interests U.S. oil companies too. They are scrambling to buy off-shore oil drilling leases around Point Reyes and at least one—and probably several—companies last month listed tracts which include the submerged island among sites they might like to buy when the U.S. Interior Deparment sells oil drilling leases along the Northern California coast in 1981.

EMPHASIS

Cordell Bank ——

Continued from Page B-1

and humpback whales that have been observed there.

This supply of nutrients, says Schmieder, apparently makes life possible at far greater depths than normal.

For safety, the divers have been able to reach only the pinnacles — 115 to 220 feet below the surface. But because of unusual light levels at this depth, they have been able to see and photograph 70 feet or more, their views blocked by swarms of fish rather than by the sediment normally found along the coast at these depths.

Until this expedition, modern civilization knew only the rough shape of the bank and that the pinnacles were of hard granite, samples of which were dredged up in 1949.

The bank was discovered in 1853 and was named after Edward Cordell of the U.S. Coast Survey, who rediscovered it in 1869 and roughly mapped its contours. The science crew quickly found that maps prepared since Cordell's time are not very accurate. They have spent hours traversing the bank taking soundings to plot the contours.

For the past 100 years, fishermen have known Cordell Bank as an ocean magnet for large birds, fish and migrating whales, a mysterious place where the currents sometimes flow backwards, and the weather is subject to sudden change especially in summer. It was the lack of information to explain this phenomena that prompted Schmieder, a physicist who also loves biology, to begin organizing the first diving expedition of the bank with a scientific purpose.

"At the highest 19-fathom level, found last year, barnacles have taken over the peaks," says Schmieder. "Further down, the granite is covered with a tremdously colorful mixture of animal life that is a foot or more thick."

Example of the variety of Marine life found on

Pastel pinks and greens, bright reds, whites and yellows make this underwater kingdom look like an Easter garden at 25 fanthoms. The encrustations are composed of enormous piles of sponges, anemones, hydroids, hydrocoral and an occasional crab or gastropod, says the scientist.

More than 200 species of marine animalsand plants have been identified among samples brought up by the divers, some previously unknown at this latitude and some from depths not previously known to support such life.

New species of algae and a rare pink snail are among the treasures brought to laboratory scientists at California universities for identification.

The first humans to see the hard granite peaks apparently were Navy frogm... '70s, ... his di... paren... Navy ... detect...

...plore undersea mountain

of Mt. Diablo if it were flattened on top, Schmieder said. It ranges from 125 feet down from the water's surface to 200 feet down.

The underwater mountain, first discovered in 1853, was once probably part of Mojave, and in the same chain as the Farallon Islands.

It's likely that 100 million years ago, movement on the San Andreas earthquake fault caused part of the land to break away, eventually forming the three- by five-mile underwater mountain.

The name, California's long-lost island, comes from the fact that the underwater mountain was first an island which could be seen from the shore and could even have been visited, Schmieder said.

It was ignored, however, for a number of reasons: cold and deep waters, strong currents, sharks and a lack of knowledge about the area.

The area is used for fishing — fleets leave from Bodega every day to fish for cod, yellow-tail and salmon.

Schmieder and his fellow explorers devoted three years of free time to the project just for themselves, he said.

The Sandia scientist got the idea for the three-year exploration when he began scuba diving as a hobby in 1974.

His group received grants from The National Geographic Society, the New York-based Explorers Club and the private Conservation and Research Foundation.

Cordell Bank surprises explorers

By Kathy O'Toole
Tribune Staff Writer

Emphasis on science

WHILE THE OIL rigs cluster off the Santa Barbara Coast in search of a new Prudhoe Bay, another expedition has been painstakingly probing the sea floor off Point Reyes for a treasure more colorful and fragile than petroleum.

The expedition has discovered an underwater biological community so rich and unusual that the National Oceanic and Atmospheric Administration is considering whether it should be a national marine sanctuary.

Such status probably would make the area off-limits to oil exploration crews and put limits on commercial fishing so that a single swipe of a drag-net could not destroy some of the 20-year-old residents of this underworld at the precipitous edge of the outer continental shelf.

The community of marine animals and plants thrives on a submerged island with steep, jagged granite ridges rising from a flat, sandy canyon that is the San Andreas fault extension.

Located 20 miles west of Point Reyes and 50 miles northwest of San Francisco, it is the northernmost outcropping of that part of the Earth's crust called the Salinian block. The island was detached from the mainland plate sometime in the last 60 million

Mystery sea mountai...

By Sheila Riley
Staff writer

LIVERMORE — Cordell Bank, an underwater mountain in an uncharted area off California's coast, was an intriguing mystery to fishermen and oceanographers.

Sandia combustion research scientist Bob Schmieder decided three years ago to learn more about "California's long-lost island," the name explorers gave to the area 50 miles northwest of the coast and about 20 miles due west of Point Reyes.

The underwater mountain, discovered in 1853, was once probably part of Mojave, and in the same chain as the Farallon Islands.

It's likely that 100 million years ago, movement on the San Andreas earthquake fault caused part of the land to break away, eventually forming the 3- by 5-mile underwater mountain.

The name, California's long-lost island, comes from the fact that the underwater mountain was first an island which could be seen from the

to fish for cod, y... salmon.

Schmieder and his f... devoted three years ... the project just for ... said.

The Sandia scienti... for the three-year ex... he began scuba divin... 1974.

His group receive... The National Geograp...

...xplore Cordell Bank

...re unknown. And he said the use of modern equipment may prove some of the earlier measurements inaccurate.

In 1949 the California Academy of Sciences performed dredging operations in the area. A bag of rocks and dried mud is all that remains of that work today.

A more mysterious exploration of Cordell Bank was made by professional divers hired by the U.S. Navy in 1969. Schmieder said he's talked to one of the divers, who lives in the Bay area, who wouldn't say what they had seen. He did say he'd made the dive between 40 and 45 times.

Schmieder said all the results were classified confidential and have never been released to the public. Virtually all he knows about the

Dr. Robert Schmieder

project is that it was related to communications.

Soviet trawlers also have been known to stop in the area of Cordell Bank for several days at a time.

After taking up diving about five years ago, Schmieder became interested in the idea of exploring the Cordell Bank while teaching physics at UC Berkeley. Several of his colleagues mentioned that there were almost no biological specimens available from the area.

"It was obvious what to do — go and get some," he said.

So about a year ago he started assembling a group of divers according to their reputation, or because they're friends and share his interest in exploration.

Only Schmieder is from Contra Costa County.

They've made several practice dives, some around the Farallon Islands south of the bank. And they've had many strategy sessions to plot out safety procedures and plans for the big dive.

"A lot of people have gotten exhausted or run out of money," Schmieder said, adding that he hoped to have at least 12 divers on the trip, but now has seven.

So there will be at least three divers on each "team," and only one will go into the water at a time. A sophisticated communications system of walkie-talkies that would have allowed the divers to communicate with one another and the boat crews on the surface has been scrapped.

"It's just too dangerous," Schmieder said. The system requires special face masks that the divers are unaccustomed to. During a practice session, one diver's mask flooded, he said.

To deal with the current around the bank, which normally flows at between one and two knots southeasterly, the crew will find the point where they want to dive and drop a small line. Then the boat will move "upstream" above the point and then drift down with the current.

Schmieder said there is a location about 20 fathoms (about 280 feet) deep where they plan to dive. Another more shallow spot may prove more interesting because there are few such locations at Cordell Bank, he said. It would be like a peak jutting up underwater. But it is so small the crew may have trouble finding it.

"This is really an exploratory dive," he said. "As far as we can tell, there have been no divers there with a scientific motivation.

"If we come back with any specimens at all, it (the dive) will be a success. We'll have an idea of what has to be done."

To date, all the costs of the now $2,000 expedition have come out of the divers' pockets. Schmieder said there have been pledges of contributions — after they return with specimens.

He acknowledged that diving has become an expensive pastime for him. "It's still a hobby. But it's a serious one. I'm not an abalone freak and I don't spear fish."

Usually, Schmieder dives along the California coast at places such as Santa Barbara, Monterey, Point Lobos or the Farallons.

He's waiting for a copy of a listing of shipwreck sites compiled by a sailor during the 1850s so he can compare it to the contemporary locations in contemporary literature.

"What interests me is going somewhere people haven't gone before. I know that sounds like 'Star Trek,' doesn't it," Schmieder said. "But I want to be doing something a little more serious than your regular Sunday diver."

Colorful coral and sponges inhabit the Cordell Bank.

Photo courtesy of Robert Schmieder

Undersea mountain reveals ancient clues

By Peter Aleshire
Times staff writer

SAN FRANCISCO — Robert Schmieder of Walnut Creek and the band of divers and scientists who have explored an undersea mountain just 50 miles northwest of San Francisco took the pair of vast blue whales that once glided under their boat in stride.

They weren't even bothered by regularly diving 300 feet below the surface, where too long a stay can lead to hallucinations and a too-rapid ascent to death.

But they did get a little unnerved when a submarine once broke the surface behind their boat and followed them for a time as they made their way back to port.

They never found out whether the sub was deliberately following them, but they're still wondering if it had something to do with the man-sized holes bored in solid rock that they discovered in the supposedly unexplored depths of the Cordell Bank.

The discovery of the four-foot-wide, eight-foot-deep holes stunned the divers. "It was like finding beer cans on Mt. Everest," Schmieder said.

He was told by unofficial sources that the Navy had been doing work in the vicinity of the bank, but Navy officials have refused to discuss the holes, he said.

Schmieder also discovered that the day after the expedition's first dive, Navy divers went into the

Please see UNDERSEA, Page 5A

Lush Sea World Rises From Obscurity

By Charles Petit
Science Correspondent

Except for fishermen and navigators of submarines, hardly anybody has heard of the Cordell Bank. But this underwater mountain north of the Farallon Islands has bewitched a 41-year-old diver named Robert W. Schmieder.

He and scuba-diving friends, lately assisted by some small private and federal grants and professional biologists, have elevated the obscure seamount to formal consideration by the National Oceanic and Atmospheric Administration as either a full-fledged National Marine Preserve or an extension of the existing preserve that embraces Point Reyes and the Farallon Islands.

Rising to within 115 feet of the surface, 20 miles from Point Reyes to the east and the Farallones to the south, Cordell Bank is, Schmieder argues, an ecological treasure house of bottom-dwelling organisms.

The waters on the landward side are typically more than 400 feet deep, and to the west plunge abruptly to an abyssal plain 10,000 feet and deeper below the sea surface. Above the bank is found some of the best fishing in Northern California. Diving birds fill the air. Gray whales and killer whales cruise over sunken peaks which, during the last ice age, probably stood above the waves as islands.

But, Schmieder fears, its living communities are vulnerable to damage by drag nets from fishing boats, and might be devastated by drilling operations should the region be opened to oil companies. Status as a marine sanctuary "would give it the protection it deserves," maintains the stocky, bearded Schmieder.

Washed by clear, cold, Pacific currents that have traveled from Japan via the Aleutians, Cordell Bank will never be visited by any but well-trained and intrepid divers.

It is "literally coated", in Schmieder's words, with a large community of algae, sponges, anemones, coral, urchins, snails, worms, and crabs. Fed by upwelling current rich in nutrients, the bank is covered with a living encrustation a foot or more thick, colored in brilliant shades of red, yellow and pink.

The bank "is probably one of the most lush underwater habitats on the Pacific Coast," agrees Gordon Chan, professor of marine zoology at the College of Marin.

Chan, one of the few professionally trained biologists ever to dive on the Cordell Bank, said that when he first saw it through his face mask a month ago during a dive organized by Schmieder, "it just opened my eyes. If you go down 120 feet off Monterey, for instance, you wouldn't see anything at all."

Species unknown to science may be found on the bank, Chan said, adding that some of the sponges he spotted are not described in any biological literature he knows. Other species, such as a variety of algae, were found on the bank farther north than they had previously been known to exist.

Further study is needed to be sure the bank's living communities deserve special protection, Chan said, "but my guess is that they are. I think it ought to be included in the Point Reyes and Farallon (preserve). It must be the basis of a very complex food web, and probably is important to both fish populations as well as marine mammals all along this coast."

But it may be an uphill battle to gain protection as a federal marine sanctuary.

Only five places have been named marine sanctuaries, including the Channel Islands off Southern California and the site of the sunken Civil War ironclad Monitor off North Carolina. Others under consideration include the waters where Humpback whales rendezvous off the Hawaiian island of Kauai, a rich and accessible diving paradise in American Samoa, and a tropical bay in Puerto Rico.

The bank is named after geographer Edward Cordell, who in 1869 became the first man to map its general contours. A nine-mile-long protuberance, it is the northernmost seamount on the West Coast, the final offshore extension of a geological formation called the Sali...

The Cordell Bank is called an ecological treasure house of bottom-dwelling organisms

Man and ...

Mo... at C...

...es mysterious island beneath Pacific

...e successful dive this year — out ... 12 attempts. In all, the group has ...ade two successful dives to the ...x-mile-long, three-mile-wide ...nk.

Why no one before him had journeyed into the frigid waters of Cor...ll Bank is obvious to Schmieder, ...hat with the physical dangers of ...ch deep dives and the area's close...ss to Tomales Bay, a breeding ...round for Great White sharks.

But why did Schmieder himself ...ke the risk — for fun?

"I had a hobby of SCUBA diving ...nd I did it for fun for the past eight ...ears," he explained. "And I was ... had done it before."

Schmieder's first dive to Cordell Bank, on Oct 20, 1978, proved many experts wrong, not only in that the feat could be accomplished, but in what the area was thought to look like.

"We found much more light (in the depths near Cordell Bank) than we expected," he said. "It wasn't terribly cold and it wasn't dark at all. The water was very clear. There were very little particulates."

Visibility in most California waters is about 20 feet and Schmieder said flashlights are required below 100 feet, because it's "pitch black."

But 150 feet from the surface at ... penetrate to greater depths, creating growth of algae and microscopic diatoms much deeper than scientists expected.

"The environment (of Cordell Bank) has been pictured in people's minds quite incorrectly," he said. "It's really forced a major revision in what the populations would be."

A brown alga normally found growing at the surface, for example, was found 150 feet down on Cordell Bank, he said.

And, he added, the group will be proposing to do a study on the distribution of the area's flora and fauna with increasing depth.

"We'd like to see if the higher light ... found on Cordell Bank, was sent to the Smithsonian Institution in Washington, D.C. for further study.

And two new genuses of algae, a brown and a red form, were sent to botanists at the University of California, Berkeley who've managed to map both plants' life cycles.

All of this excites Schmieder and his crew, which includes biology buffs and graduate students who come along for the excitement of discovery.

The group this year received $3,000 in grants from the National Geographic Society and the Conservation and Research Foundation for the expedition, but the project was

Cordell Bank Dive Was Well Worth...

WALNUT CREEK — For Bob Schmieder and diving crew, the 19th time was the charm.

That's how many times the Sandia Laboratory physicist and amateur diver had scheduled an exploratory dive to the submerged Cordell Bank about 20 miles west of Pt. Reyes.

"We finally had a break in the weather," Schmieder said after the all-day Friday expedition to begin charting and specimen collecting on the virtually unstudied shoal.

He started trying to schedule dives in mid-summer, but was stymied by rain, wind and fog. From a crew of 15 divers interested in the project, only five made the trip last week.

"Scientifically, so far it's been very rewarding," Schmieder said.

Dozens of specimens of algae and invertebrates were gathered and two of the divers, members of the UC Berkeley botany department, took them back to labs for identification. At least one new genus of algae was reported, Schmieder said, as well as several specimens of algae never discovered at such depths.

Tribune photo by Howard Erker

Alth... make ... Schmi... weathe... vent an... year.

The f... of Cord... when G... ticed h... veered ... he coul...

INTRODUCTION

Underwater Island

Thick fog now. Just my luck. Am coming in right now as I write these words, feeling for the Cordell Bank ...

Captain Irons Pendleton, the hero of Sterling Hayden's epic novel *Voyage*, was tense. Twenty or thirty miles out in the ocean west of Pt. Reyes, he groped his way through the fog toward the Golden Gate. From the Coast Pilot he knew that somewhere nearby there was a mountain underwater.

Armed the lead with tallow and let it plunge into the seething wake. Two hundred fathoms of wire ran out ...

Pendleton waited hopefully for the relatively shallow water to tell him he was where he wanted to be. Somewhere down there were hard rocks, with barnacles and red slimy things, but all that was really important to him that night was that it was *shallow*. No one had ever seen this place, and he could only imagine what it looked like: a high and mighty rise above a flat plain, with many things living on it ... like an island ... only underwater ...

The existence of this underwater island was known since shortly after the California gold rush. But Pendleton, like hundreds of real 19th century mariners, could not know how exotic it really is. They could not know that it presents one of the visually most spectacular underwater sights in the world. Or that it harbors an extraordinarily lush community of algae and invertebrates that in turn supports a huge number of larger animals, including fish, seabirds, and marine mammals.

Cordell Bank lies within sight of San Francisco Bay, yet it has been mantled in mystery, and remains so today [Garretson, 1978]. Indeed, until a few years ago, this exquisite site had never been seen by humans, even though it lies only a few hours by boat from a major metropolitan area. Only recently have we begun to explore its erosion-remnant topography, to catalogue the little creatures that inhabit its craggy peaks, to observe the great birds and mammals that frequent the surface waters, and to appreciate its pristine isolation. And although we are at the very beginning of our understanding of the complex processes at work on the Bank, we have learned enough to give a rough picture of its ecology and biogeography. This book summarizes our present understanding of the plants and animals that live on this underwater island, and how they all make a living together.

Finding It

First, you have to get to California. San Francisco is a good place to have breakfast before your expedition. Perhaps a leisurely drive across the Golden Gate Bridge and up the Pacific Coast Highway.

Without a boat, the easiest way to find Cordell Bank is to stand at the lighthouse at Pt. Reyes; and look due west, out over the ocean. If the day is clear, the horizon will appear roughly 37 km (20 nautical miles) away. Right at the horizon is an underwater mountain, a rocky outcrop 15 km long and 8 km wide [NOAA, 1982]. Over most of it the water is roughly 60 m deep, but there are a few ridges and pinnacles that rise to within 40 m of the surface. The entire Bank, especially the shallowest parts, is teeming with life.

Now look to your left, due south. Exactly the same distance away as Cordell Bank (37 km) is the Southeast Farallon Island the largest of several granitic islands, some totally submerged. Now turn completely around and look due north. Again, exactly 37 km away, is the rocky head at Bodega Bay.

From Cordell Bank, it is 100 km to the Golden Gate Bridge. But there's only one spot in San Francisco that you can see from Cordell Bank: the top of Mt. Davidson, 285 m high.

Cordell Bank lies on the very edge of the continental shelf like a castle tower at the edge of a cliff. From its rocky heights, the ocean bottom falls off extremely rapidly to abyssal depths: in a distance less than the width of the Bank, the water is more than 1.5 km deep [NOAA/NOS, 1974]. Between Pt. Reyes and the Bank is a broad, relatively flat basin filled with muddy sediment, about 120 m deep. This basin is the graveyard of countless tiny organisms such as foraminifera that attain partial immortality by synthesizing hard durable shells.

The underwater island, Cordell Bank, lies on the west coast of North America, 100 km from San Francisco. In this azimuthal equidistant projection, the Bank is placed at the center; the World's landmasses are distorted so that circles centered on the Bank represent points equally distant from the center. The outer circle is the antipode, 10,800 nautical miles distant, and the scale is linear. Therefore, a radial line from the Bank to any point gives the shortest (great circle) path and the proportional distance. (Inset) The nearest landfall to Cordell Bank is Pt. Reyes, 20 nm distant. Due to the strategic position of this underwater island at the northern entrance to San Francisco Bay, it became a well-known point of reference for mariners.

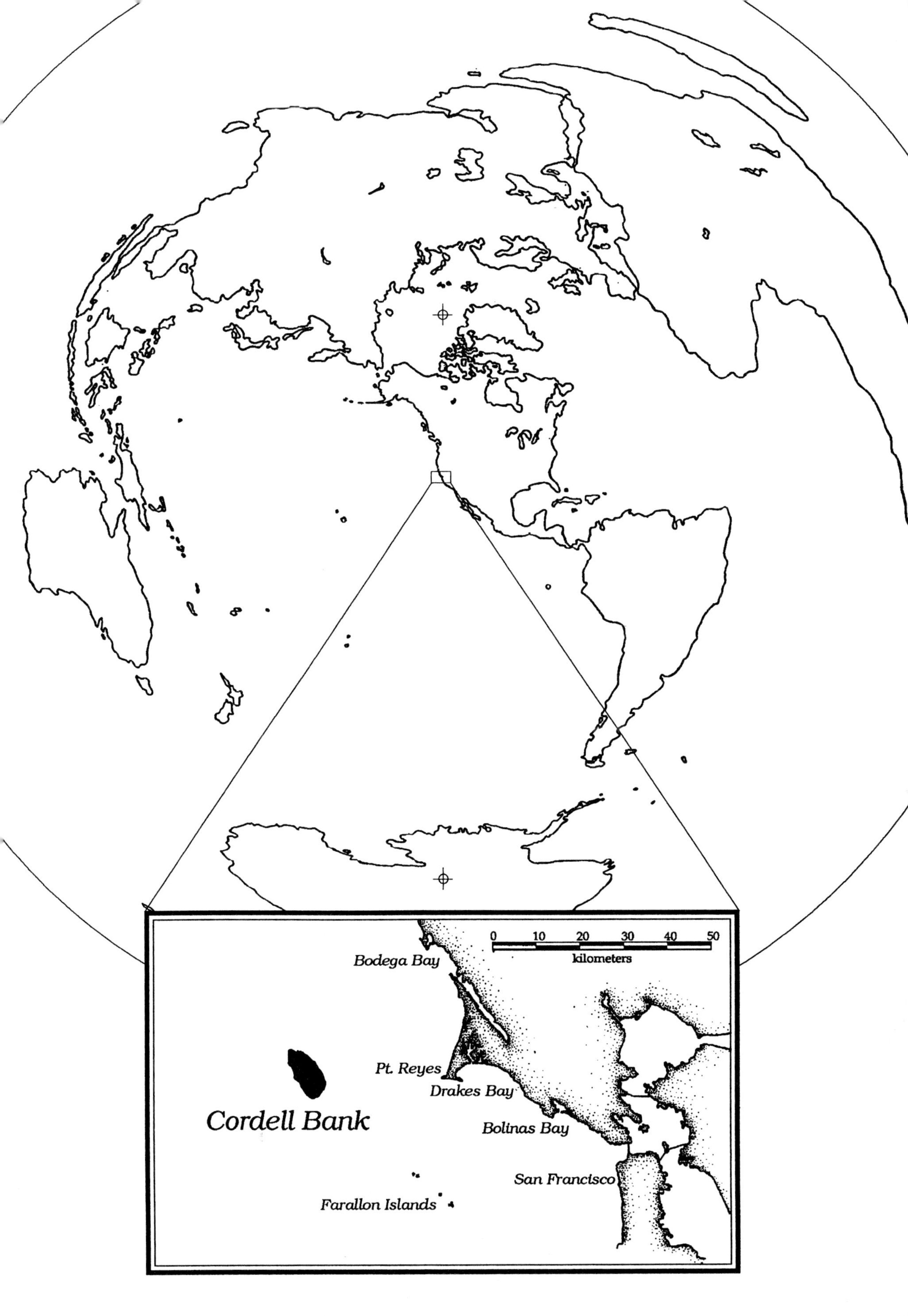
0
10
20
30
40
50
kilometers
Bodega Bay
Pt. Reyes
Drakes Bay
Cordell Bank
Bolinas Bay
San Francisco
Farallon Islands

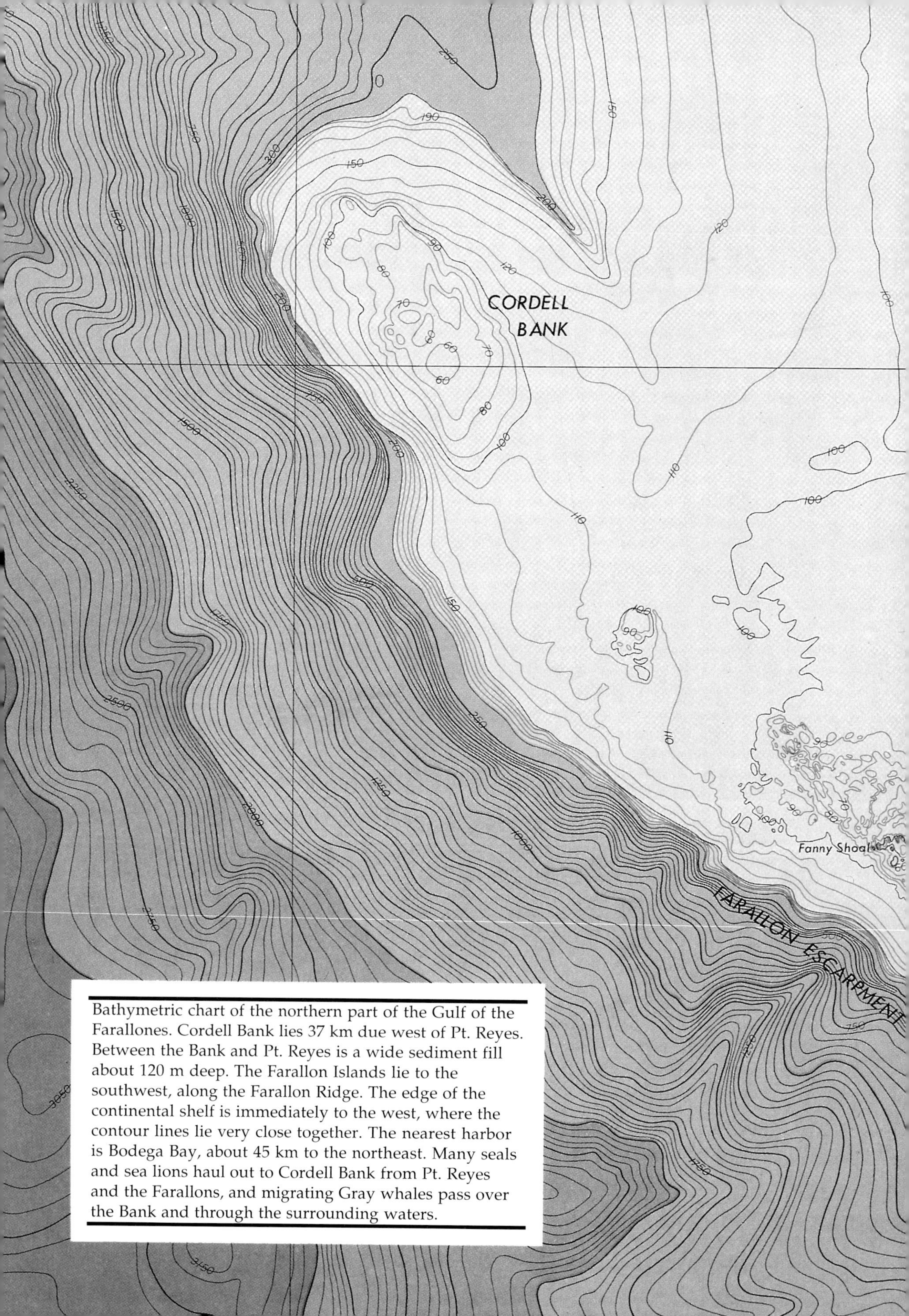

Bathymetric chart of the northern part of the Gulf of the Farallones. Cordell Bank lies 37 km due west of Pt. Reyes. Between the Bank and Pt. Reyes is a wide sediment fill about 120 m deep. The Farallon Islands lie to the southwest, along the Farallon Ridge. The edge of the continental shelf is immediately to the west, where the contour lines lie very close together. The nearest harbor is Bodega Bay, about 45 km to the northeast. Many seals and sea lions haul out to Cordell Bank from Pt. Reyes and the Farallons, and migrating Gray whales pass over the Bank and through the surrounding waters.

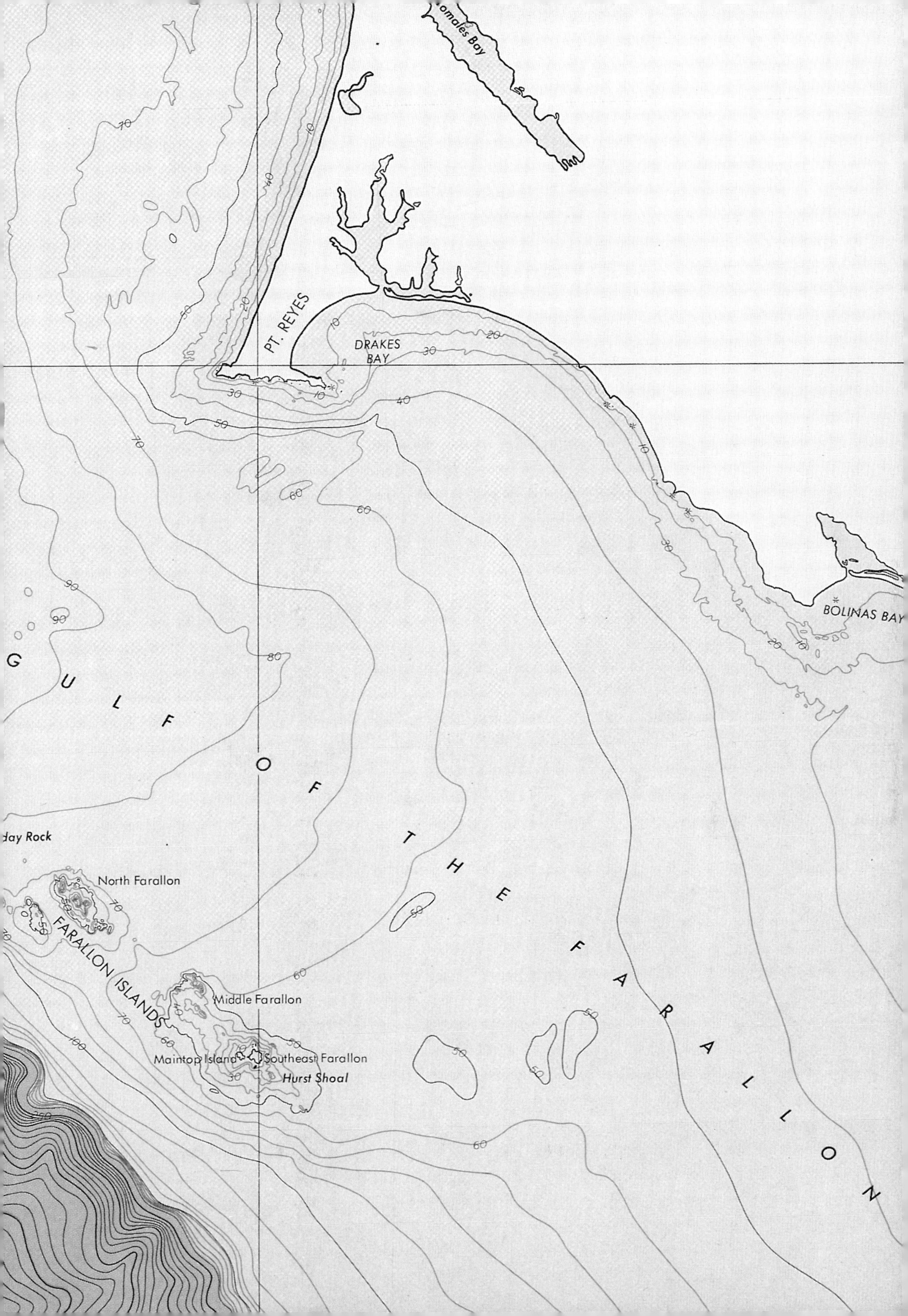

omales Bay
PT. REYES
DRAKES
BAY
BOLINAS BAY
G U L F O F T H E F A R A L L O N
day Rock
North Farallon
FARALLON ISLANDS
Middle Farallon
Maintop Island
Southeast Farallon
Hurst Shoal

A Great Place to Live

It's the water that allows Cordell Bank to burst forth with living things. Lying sufficiently far offshore to be relatively free from mainland sediment, the Bank is continually washed with relatively clean, cold, oceanic water [Reid, 1959]. Upwelling of deep, nutrient-rich water all along the northern California coast provides organic materials utilized by growing phytoplankton. Because of the clarity of the water, sunlight is able to penetrate significantly deeper than in the turbid, nearshore waters, bringing with it energy to drive photosynthesis. Zooplankton consume the phytoplankton, and larger animals consume smaller ones. The irregular formation of eddies and jets of flowing water help carry food to the Bank and carry debris away from it.

With plenty of food, the animals grow luxuriantly: the tops of the ridges and pinnacles are smothered in a fantastic jumble of sponges, anemones, hydrocorals, hydroids, tunicates, barnacles, crabs, worms, scallops, snails, chitons, starfish, and red, brown, and green algae, to say nothing of the entoprocts, ectoprocts, brachiopods, copepods, amphipods, ostracods, scaphopods, isopods, nemerteans, rotifers, nematodes, gastrotrichs, kinorhynchs, and tardigrades! Circulating placidly above this biological circus are large numbers of fish. These fish eat little creatures such as snails, worms, and smaller fish, and in turn are eaten by bigger fish and by the birds and mammals (including man) that reach down for them from the surface.

There are many more animals than plants at Cordell Bank. This is true whether you count the number of plant and animal species, or measure the biomass per unit area, or estimate the percentage of the cover from underwater photographs. This faunal preponderance is due to the inability of plants to photosynthesize in the low light level at great depths, and the fact that the animals, able to utilize high energy food supplies, are much more successful in generating biomass than are the plants.

A computer-generated sonar image of Cordell Bank. This image was made from approximately six million soundings, each a measurement of the average depth of an area on the Bank roughly 4 m square. The data were automatically recorded onboard the survey ship and later processed into images. Shadowing was artificially added by simulating sunlight at a low angle from the northwest, enhancing the perception of depth. Images such as this show for the first time the topography of the Bank, and demonstrate that its surface was sculpted by fluvial erosion.

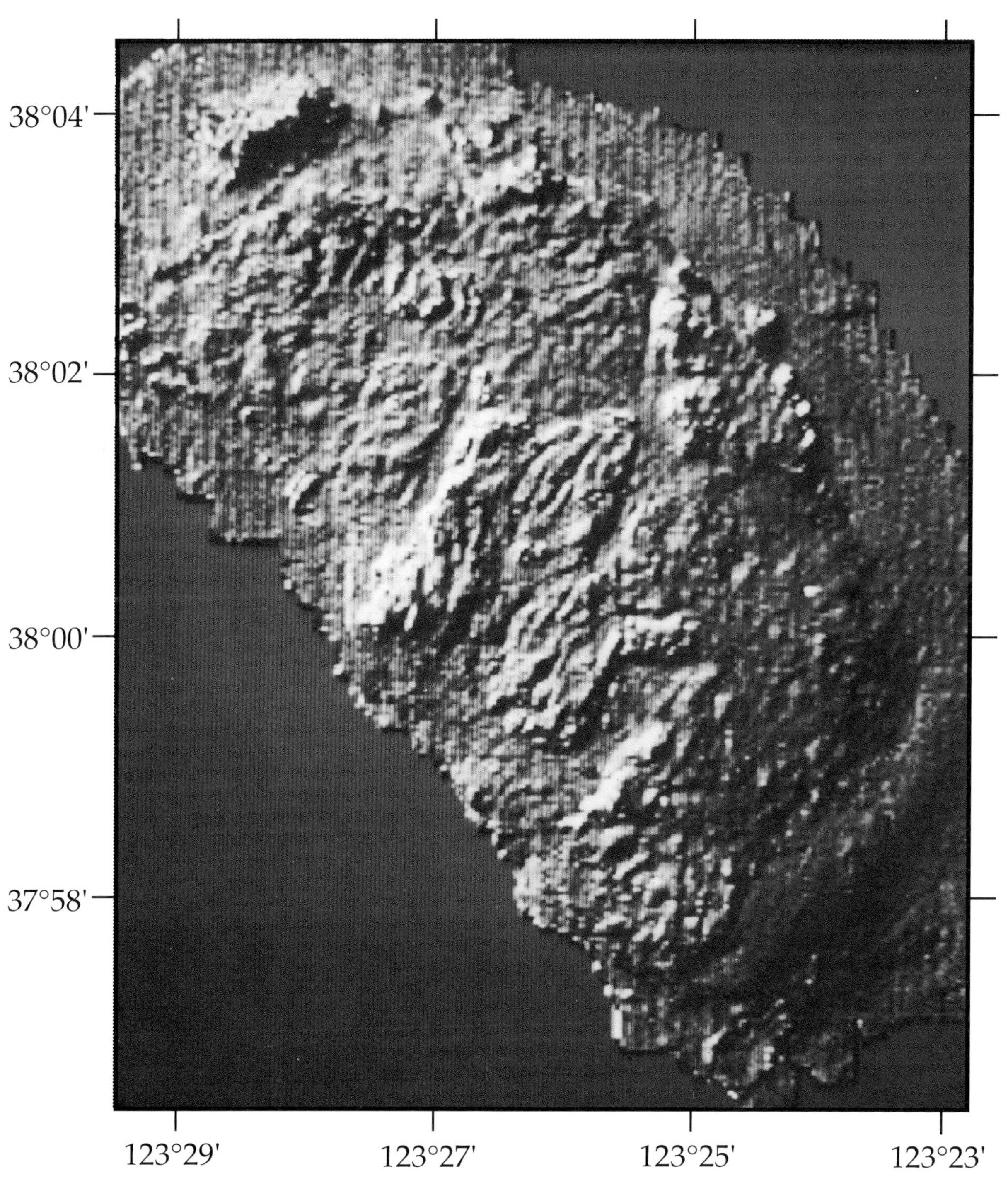
38°04'
38°02'
38°00'
37°58'
123°29'
123°27'
123°25'
123°23'

More Than Crowded

One consequence of full stomachs, however, is fecundity. The animals reproduce with abandon—and find themselves competing with their offspring for the same piece of ground. The principal limitation to the numbers of individuals is available substrate—a place to be. Forced to find a foothold or be washed off the rocks, the organisms pile on top of one another, snails on algae on sponges on hydrocoral, all reaching for open space to continue searching the water for more food. The result is a community in which commensalism (one organism living with another that is unaffected), parasitism (one organism living on and at the expense of a host), and symbiosis (two organisms living together for mutual benefit) are prevalent, and in which a relatively small number of species accounts for the great majority of the biomass.

(Above left) The author exploring the cover on Cordell Bank, at a depth of 45 m. The great majority of the cover is sponges, plus a few other animals. Practically no plants are in evidence.
(Above) The top of a shallow pinnacle, about 40 m below the surface. The pinnacle is crowded with overlapping clusters of animals, plus a small amount of algae. Dominating the scene are large white sponges, anemones, tunicates, and large colonies of the pink California hydrocoral. Wandering around looking for prey and grazing on detritus are a few snails and crabs, while hundreds of fish of all sizes circulate above, waiting for an opportunity to snap up an unwary creature. Off the edge of the pinnacle, the water is deep—and dark. The walls of the pinnacle are practically vertical, and any organism that is unlucky enough to lose its grip will float away to its death or tumble down to the sediment deposits far below.

The algae at these depths, in contrast, have to cope with the low absolute light level; photosynthesis proceeds much slower than in shallower communities. In reality, compared with nearshore areas of comparable depth, the Bank supports a relatively rich growth of algae: many species are found along the coast only at much lesser depths [Silva and Moe, 1978]. This differential is attributable to the fact that the light level is *relatively* high, due in turn to the clear water on the Bank relative to comparable nearshore areas.

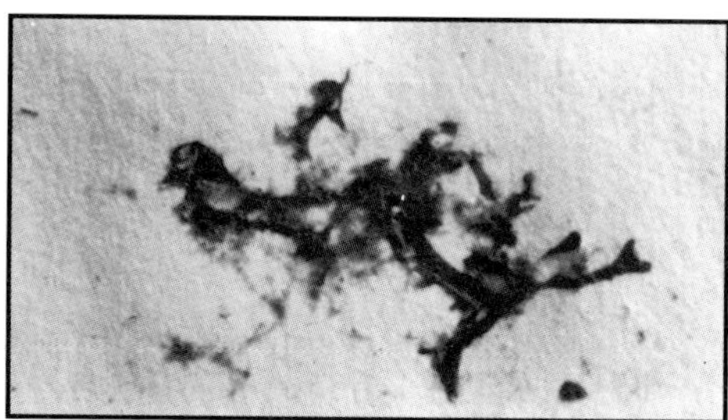

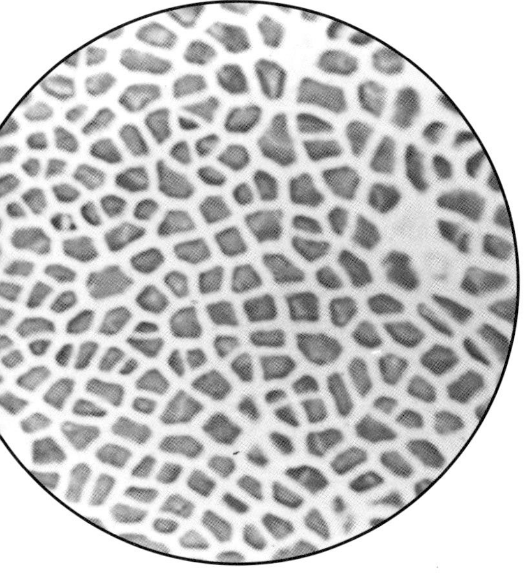

(Above) A mounted specimen of an undescribed alga in the family Delesseriaceae (magnification 1.8x). (Left) Photomicrograph of a female plant, showing the characteristic reticulated cell pattern (magnification 250x). (Below) Male plant (magnification 100x). Although this alga appears rather unimposing, discoveries such as this support the idea that Cordell Bank is somewhat isolated: some organisms live there and nowhere else.

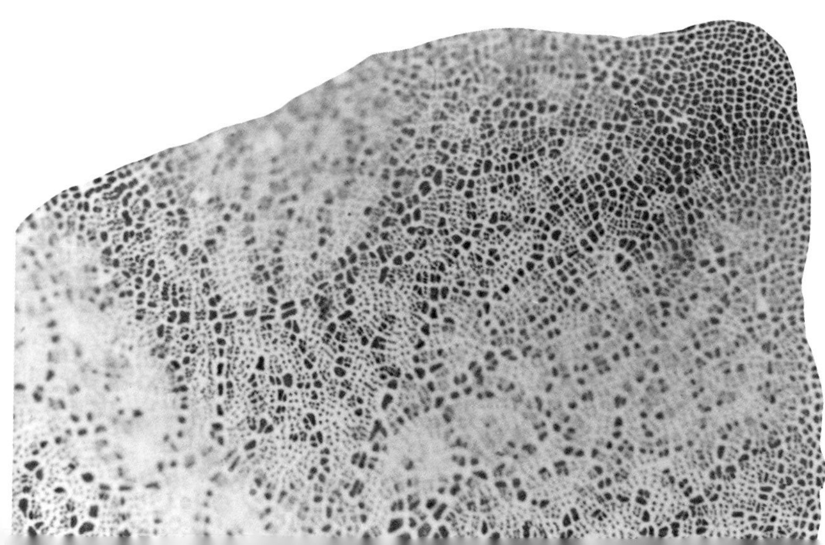

New and Different

Several previously unknown plants and animals have been discovered at Cordell Bank. In particular, two new genera and at least one new species of algae were collected by divers and brought into culture by specialists at the University of California at Berkeley [Silva and Moe, 1978]. Somewhat to our surprise, some algae that were not visible in the collections grew into large plants when cultured.

In the Animal Kingdom, there are several previously undescribed sponges living on the Bank. A scallop with two patterns of microsculpture on its valves may actually be two different subspecies [Roth, 1978]. The shells of at least two previously undescribed snails were found in the sediment [McLean, 1985]. In a different case, two previously described species of a pink snail were found to be two forms of a single species, *Pedicularia californica* [Schmieder, 1980].

Missing

There are, however, numerous plants and animals familiar from the mainland seashore that are conspicuously absent from Cordell Bank. There are no kelp forests, very little green algae, and no surf grass. None of the large clams (like the razor, gaper, geoduck, or jackknife) that are found in estuarine mud and sandy beaches are found here. There are no gooseneck barnacles or mussels that form familiar carpets on mainland shore rocks. There are no oysters, no sand dollars, no sea pens or sea pansies, no sea fans, no sea hares, no urechis worms, no moray eels, practically no abalones, and no lobsters. Furthermore, there are no sea otters or walruses. And, believe it or not, there may be very few great white sharks!

Deep Dwellers

If this list makes it appear that Cordell Bank is impoverished, don't forget that many species flourishing on the Bank are seldom if ever seen near the shore. These include many species of red algae, sponges, anemones, worms, gastropods, ectoprocts, brittle stars, and brachiopods. Many of these species are little-known inshore; the best place to find them is on a rocky seamount like Cordell Bank. Although it is true that the number of species on a geographically limited area such as a bank top will always be less than found on the adjacent mainland, we have found that the favorable conditions existing at Cordell Bank produce a community that, for its depth and isolation, is surprisingly rich and varied.

One expression of this richness and the unusual conditions on the Bank that produce it is the occurrence of many species deeper or farther north or farther south than ever before known (see page 81). More than 30 species have been collected at Cordell Bank deeper than ever before recorded. In fact, several species (including the red-striped acorn barnacle, a turban snail, and a kelp) are normally considered strictly intertidal; their occurrence on the Bank is significant evidence that Cordell Bank has some component characteristic of the intertidal.

Fossils

Although no fossils have yet been specifically identified from Cordell Bank, we can be reasonably confident that some very old materials do exist there. Mixed in with the sediment deposits are shells of mollusks and other calcareous debris from animals that lived long ago. The constant mixing by the currents of the top layers of the sediment suggests that some of the oldest remnants might be mixed in with recent materials. Fragments buried deeper than the mixing layer can be expected to be older. Identification of which shells are very old is a job for a paleontologist using sophisticated analytical techniques such as isotope or amino acid analysis.

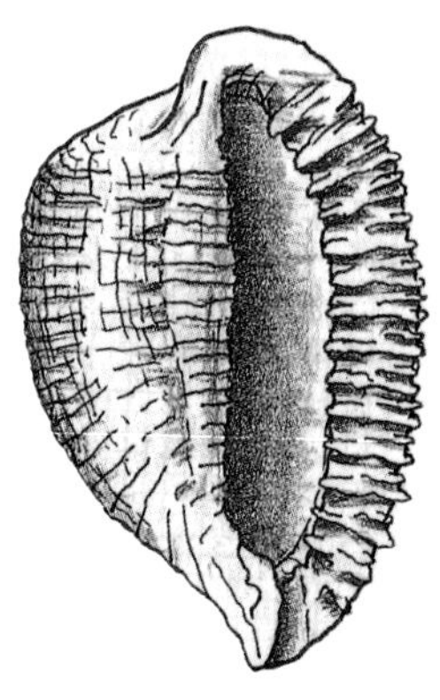

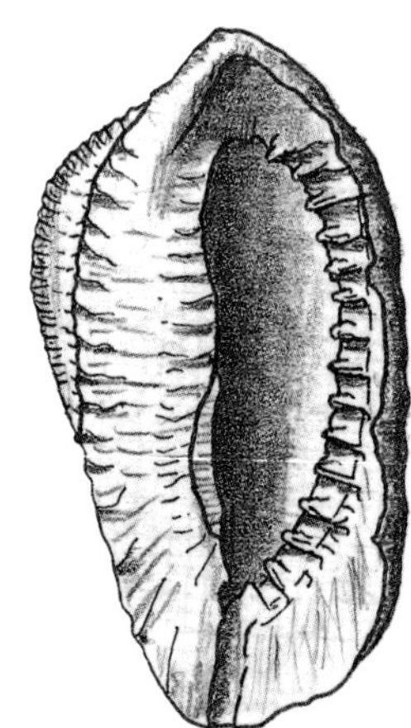

Two versions of the same snail, *Pedicularia californica*. A continuous sequence of morphologies among collections of these showed that they were all the same species. For other pictures of this snail, see pages 48-49 and Plate 46.

A selection of tiny gastropods (and two bivalves) from Cordell Bank, as seen by a scanning electron microscope. All these shells are less than 2 mm in length. It is possible that some of these could be sufficiently old to call them "fossils."

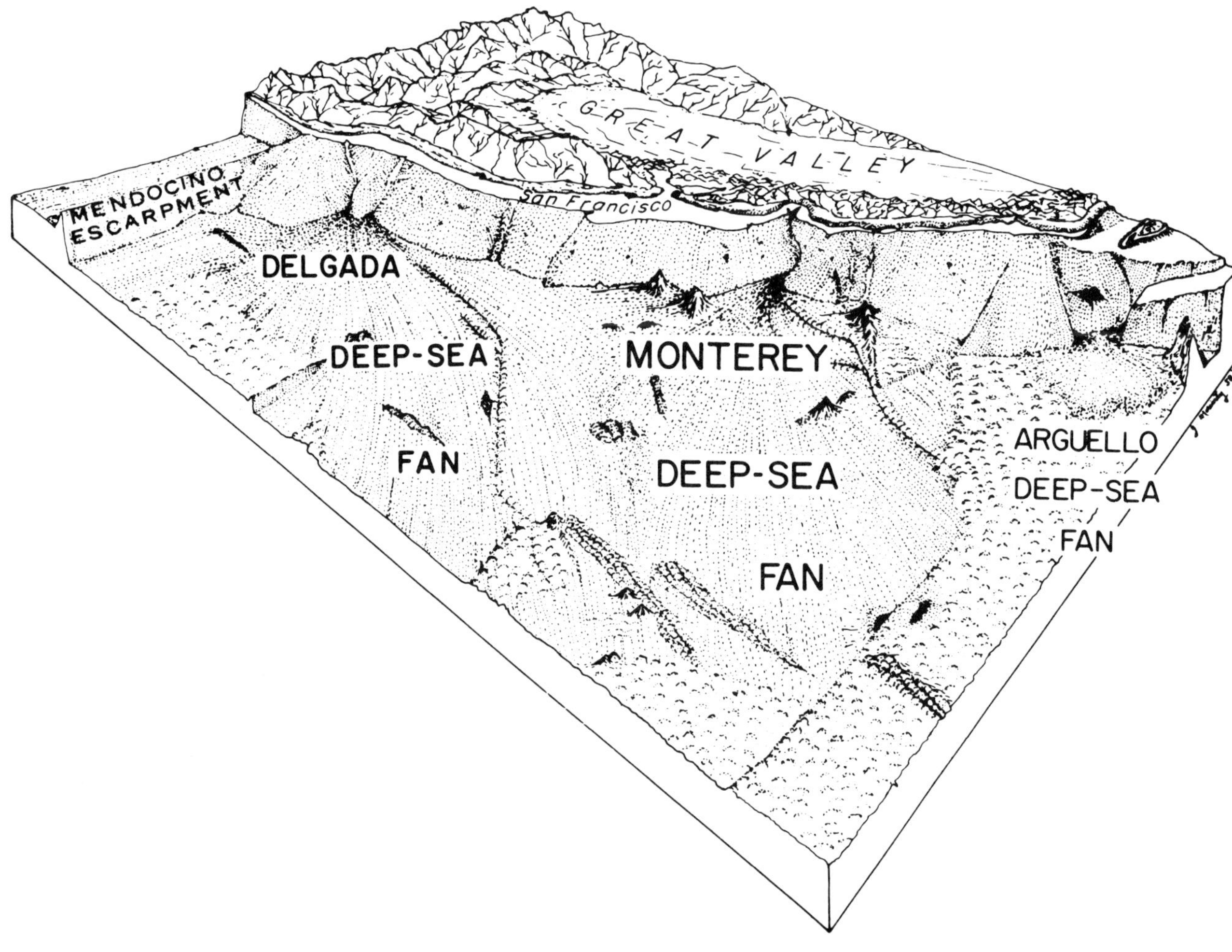

Perspective view of the continental shelf and sea floor of northern California. The abyssal plain shows evidence of parallel fractures associated with plate motions, but obscuring these are several overlapping sediment fans. The actual edge of the continent lies offshore some 15-30 km. Just under the "a" in the word "Francisco" is a small fold in the shelf. This is due to protrusion of the Salinian block as it slides northwest along the San Andreas fault. The inside of the fold forms a canyon, the Bodega Canyon. Cordell Bank lies within this fold.

GEOLOGIC HISTORY

Moving Mountains

Cordell Bank originated in southern California, perhaps even from as far away as Baja California. It was created as part of the Sierra Nevada, and pushed to where it is now by the Pacific Plate. How it was formed and got to where it is now is part of the fascinating story of the geologic ups and downs of this turbulent State. And we shall see that the biological history of the Bank is inextricably linked with its geologic history.

For this story to be fully appreciated, a few words about plate tectonics are appropriate. The outer crust of the Earth is covered with about a dozen enormous, interlocking, relatively rigid plates [Dewey, 1972]. These plates are created by extrusion of deep material along irregular borders, most of which lie in the middle of the oceans (the "mid-ocean rises"). The plates grow and move away from their sources, and are destroyed by being overridden by other plates. Some of the plates (such as the North American, South American, African, Eurasian, and Australian) carry continents that float along on them, while others (such as the Pacific, Nazca, and Cocos) are purely oceanic, without a major landmass. Where a continental plate meets an oceanic plate, the oceanic plate usually is overridden by the less dense continental plate.

This is what happened in California. The North American Plate was moving toward the west, carrying the North American continent on its leading edge, while an oceanic plate now long gone was moving toward the east. The resulting collision did terrible and wonderful things: built the Sierra Nevada, smashed huge parts of the State, and stole pieces of what is now Mexico.

Collisions

The rocks from which Cordell Bank is made were formed in this collision during the middle of the Cretaceous period, about 100 million years ago. As North America overrode the oceanic plate, the pressure generated considerable heat, producing volcanoes and forming granitic batholiths (large intrusive masses of rock) several km below the surface. Under this pressure, the ancient Sierra Nevada was uplifted, carrying the batholiths with it. By the beginning of the late Cretaceous, the magma cooled and crystallized, forming the rocks of which Cordell Bank is composed [Howard, 1979]. Isotopic analyses [Mattinson, 1982] of samples of the granite bedrock from Cordell Bank have yielded its age: 93 million years.

As the North American continental plate continued to move and grow toward the west, it completely consumed the ancient oceanic plate. Offshore, two other plates were growing: from a spreading center in the Pacific Ocean, the Farallon Plate was growing toward the southeast; the Pacific Plate was growing in the opposite direction, toward the northwest. As the North American and Farallon Plates collided, the former overrode the latter. The collision scraped ocean-floor sediments into mountains that are now called the Coast Range. The edge of the continent reached the spreading center about 30 million years ago, and suddenly it was no longer a head-on collision. Rather, the North American and Pacific Plates were both moving in roughly the same direction, but not at exactly the same rates. Compression suddenly became tension, and California began to lose its edge.

Salinia

By a quirk in the shape of the edge of the continent, the motion of the Pacific Plate was almost exactly parallel to the California coast. As a result the Pacific Plate began shaving thin slices off the edge of California and carrying them northwest along the coast. A huge and complex series of roughly parallel faults was created which is now known as the San Andreas Fault Zone [Anderson, 1971]. From the landmass that eventually would be called Baja California and southern California, a chunk was detached and moved hundreds of km. This chunk now is the strip of land that runs from Santa Barbara to San Francisco. Included in this complex strip is a huge block of hard granite called the Salinian Block, or Salinia (named after Salinas, where it outcrops) [Ross, 1978].

Thus, Cordell Bank was formed in a complex sequence of steps: first, the Salinian granite was formed deep underground about 93 million years ago. Then it was uplifted, exposed and eroded. Finally it was transported to its present position. Much of Salinia is now eroded away, and the remainder is mostly deeply buried under more recent sedimentary rocks. One high spot that pokes up out of this sediment is Cordell Bank.

The Salinian block has moved northwestward at an average rate of not quite two meters per century [Page, 1982]. At the time of our earliest human ancestors (about 3 million years ago), Cordell Bank was about 60 km south of its present position relative to the northern California coast (i.e., south of where Half Moon Bay is today). As humans began to populate North America, about 12,000 years ago, Cordell Bank was slightly more than 200 m south of its present position. Since Columbus' arrival in the New World, the Bank has moved about 10 m. And it's still moving, a few centimeters per year.

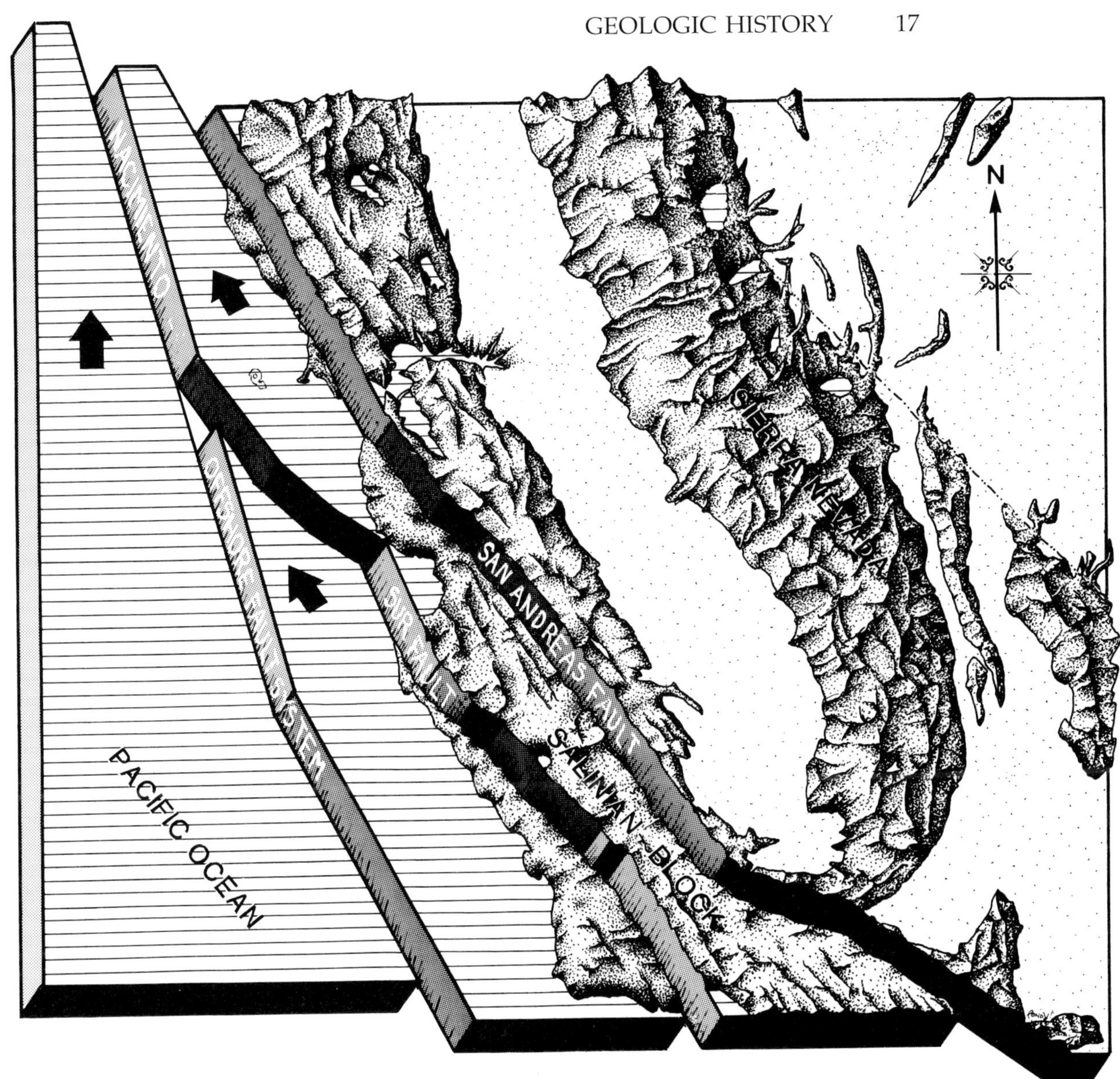

Diagram showing the motion of the blocks off northern California. This drawing shows that the shear motion of the plates is causing slices of the continent to be displaced to the northwest. One of these slices is called the Salinian Block, a chunk of granitic material that was originally a southern extension of the Sierra Nevada. Cordell Bank is a high place on this block. The block has moved hundreds of km to the northwest, carrying Cordell Bank, as well as the Farallon Islands, the Pt. Reyes peninsula, Santa Cruz, and Monterey Bay with it, at the rate of a few centimeters per year. This drawing was constructed from contours, hence the details of the folds in the mountain ranges are accurate.

Long Lost Island

Between 20,000 and 15,000 years ago, the Earth experienced a massive cold spell. Glaciers covered much of North America and Europe, and so much water was locked up in these glaciers that sea level was approximately 110 m below its present level [Milliman and Emery, 1968]. But since most of Cordell Bank lies shallower than 60 m, the Bank must have stuck out of the water by some 40 m or more. In other words, Cordell Bank was a real island! In fact, it would have been the only island of any consequence on the northern California coast, since the Farallons had emerged completely to form a peninsula connected with the mainland. The shallowest points on the island would have reached at least 60 m above the water level, and probably much more because their fragile tips have been significantly eroded.

Variation of the level of the ocean in time, and its correlation with the depths of terraces observed on Cordell Bank. Below, the sea level over the past 600,000 years is plotted, showing roughly regular oscillations driven by climate changes. At right is a simplified average cross section of Cordell Bank. At far right, the fractional area of the Bank as a function of depth is plotted; a peak on this curve indicates a relatively flat region, or terrace. The terrace depths agree reasonably well with the deepest minima of the sea level curve, or *low stillstands*. This taken as evidence that the terraces are relictual wave-cut platforms. The actual profile of the Bank contains much more structure than these smoothed plots: it preserves a detailed record of the rise and fall of the sea level. Unfolding the profile to yield the sea level curve is complicated because several stillstands overlapped at the same depth and steep slopes can catastrophically collapse.

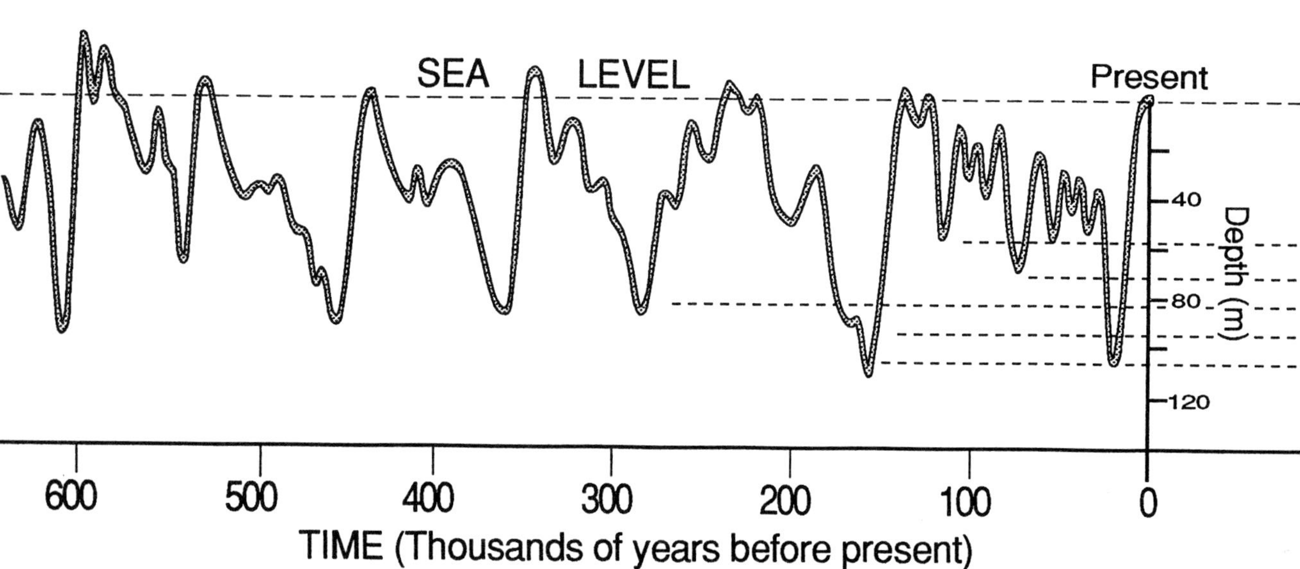

Sketch looking toward San Francisco from about 150 km to the northwest and at an elevation of about 9 km, seen as if the ocean were drained away. San Francisco itself is over the horizon and is hidden. In the foreground, Cordell Bank rises as a flat plateau above a sedimentary plain. The continental shelf falls away to the west and south. In the distance at left is the Pt. Reyes head. About the same distance away on the right is the Southeast Farallon Island. This sketch was generated from contour charts, hence is accurate in its details of the ridges, pinnacles, and valleys.

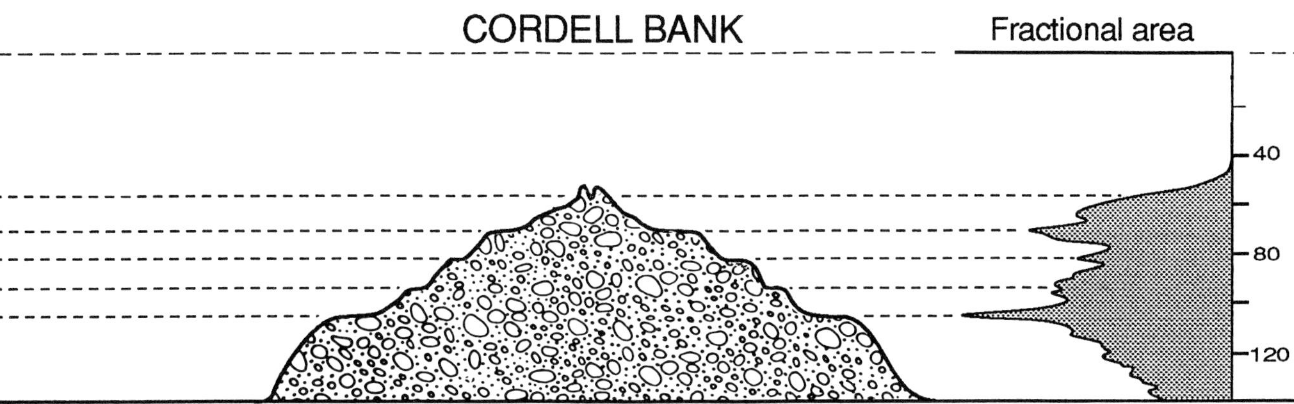

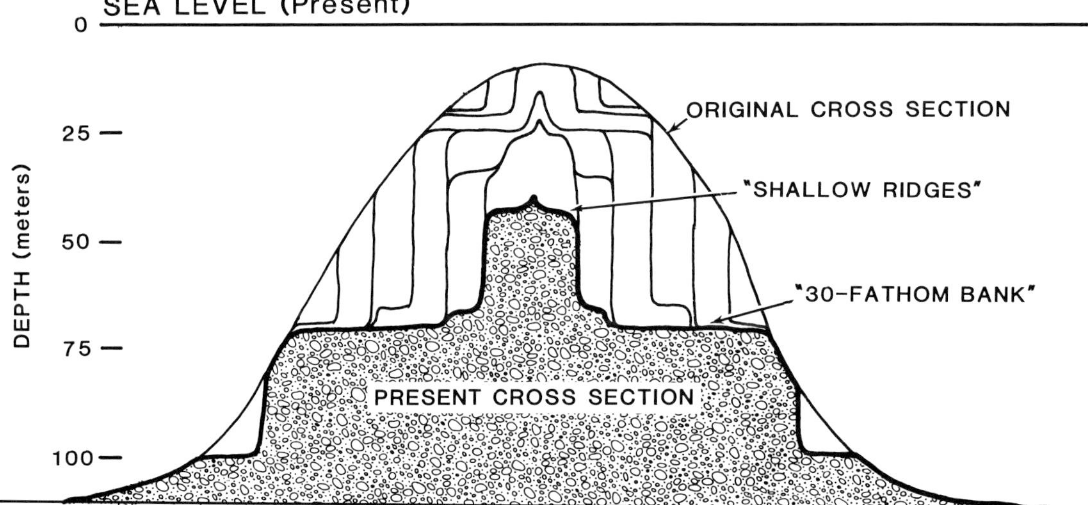

Proposed explanation for the present profile of Cordell Bank, based on the hypothesis that the topography was created by surf erosion during the rise and fall of sea level. The surf line acts like a saw, cutting into the base of the rocks within a narrow vertical zone. As this cutting edge rises and falls, the edge of the Bank is removed and the profile is modified. In this diagram, the platforms cut during successive stillstands are emphasized. The initial profile of the Bank is conjectured, but probably is unimportant since it was all removed long before the present profile began to take shape. The "shallow ridges"at present are almost entirely lost. The largest area of the Bank is at about 70 m; this flat top is due to the overlap of many stillstands near this depth which planed the top away. The stillstand which occurred about 18,000 years B. P. (before present) created the single notch at 100 m depth, clearly seen as a narrow terrace.

There is clear evidence of the fact that Cordell Bank was once an emerged island: carved into the flank of the Bank is a notch, or terrace, a relatively flat interval bounded by relatively steep margins. This terrace was cut by wave action at the surf line. Computer models of this process demonstrate how efficient the physical process is, and how characteristic the notch is of a sea level low stillstand.

When the glaciers began to melt at the end of the ice age, the sea level rose again [West, 1969; Emery, 1960]. This process was repeated many times over the Pleistocene (the past million years). Small variations in the Earth's orbit produced climatic variations, and much of the ocean's water was frozen into landward glaciers, then released again in gigantic melts. As the surf line oscillated

Computer image of a small plateau at the center of Cordell Bank. This plateau can be seen in the center of the figure on page 7. It is about 1/2 km across (left to right) and lies at a depth of about 45 m. The depth visualization has been enhanced by adding artificial light coming in at a low angle from the north. The plateau is extraordinarily flat—it varies by only 1 m or so over its width. This flatness is evidence that it was planed off by surf erosion during a stillstand. A mystery is the flat-bottomed trough that almost entirely surrounds the plateau; perhaps it is a sediment-filled drainage channel that was cut when the surf level was slightly lower than the plateau.

up and down across the Bank, a series of notches was cut. The detailed profile of the Bank records the cumulative effects of this cutting. In principle, it is possible to use a computer to unravel the present profile of the Bank backward in time, thereby determining the variation of sea level over the past million years as well as determining what it must have looked like before the erosion started tearing it down.

On the landward side of the Bank, the terraces can be traced almost the entire north/south length of the Bank. Surprisingly, they show clearly that the Bank is tilted, by about 0.1°; the south end is some 20 m higher than the north end. The direction of this tilt indicates uplift of the southern end. If the entire tilt took place since the last submergence, the rate would have been about 1 mm per year, a relatively modest figure compared to the horizontal motion. There has been relatively little overall uplift of the Bank.

Geobiology

All this geologic history is crucial to attempts to account for the biota (all the plants and animals) on the Bank, since many of the major taxonomic groups have evolved during the period over which the structure of the Bank evolved. For instance, most of the predatory gastropods (snails) now found there evolved during the Cretaceous [Taylor, 1981]. The suitability of various plants and animals for the conditions that now exist on Cordell Bank was determined by the coevolution of the biota and the landscape. In this region, one cannot talk of evolutionary biology without also talking of evolutionary geology [Valentine, 1973].

We should not forget that the topography of the Bank was radically altered during submergence. As an island, it probably wasn't nearly as flat on top as we find it today. Possibly it looked much the way South Farallon Island looks now: steep in places, barren, water worn, and smothered with birds.

The events associated with submergence are also important to attempts to account for the present biota on the Bank. During the glacial period there was a vigorous intertidal community. As the waters rose, many species found themselves in water that was too deep and became locally extinct. But others may have been just able to hold on, perhaps hunkering down and hanging on, unable to thrive, perhaps unable to reproduce normally, but also unable to die. These relicts are now stranded, and should provide valuable clues to the effects of stress on the biota and the mechanisms of extinction.

Even the tilt of the Bank may be important to understanding the evolution of the biota there. Currently, the shallowest pinnacles at the north and south ends lie at nearly the same depth (about 40 m). But the northern pinnacles must have been some 20 m higher than the southern ones when submergence commenced. Since these pinnacles now reach into the photic zone (the zone in which there is sufficient light to permit photosynthesis), this depth difference would have been significant in determining the date of the last intertidal populations, and therefore the age of any relictual organisms.

Computer images of small areas on Cordell Bank. The finest detail in these pictures is about 4x4 m square. The tips of the pinnacles are covered with dense biological communities, while between them sediment collects in traps and is slowly ground into gravel, then sand.

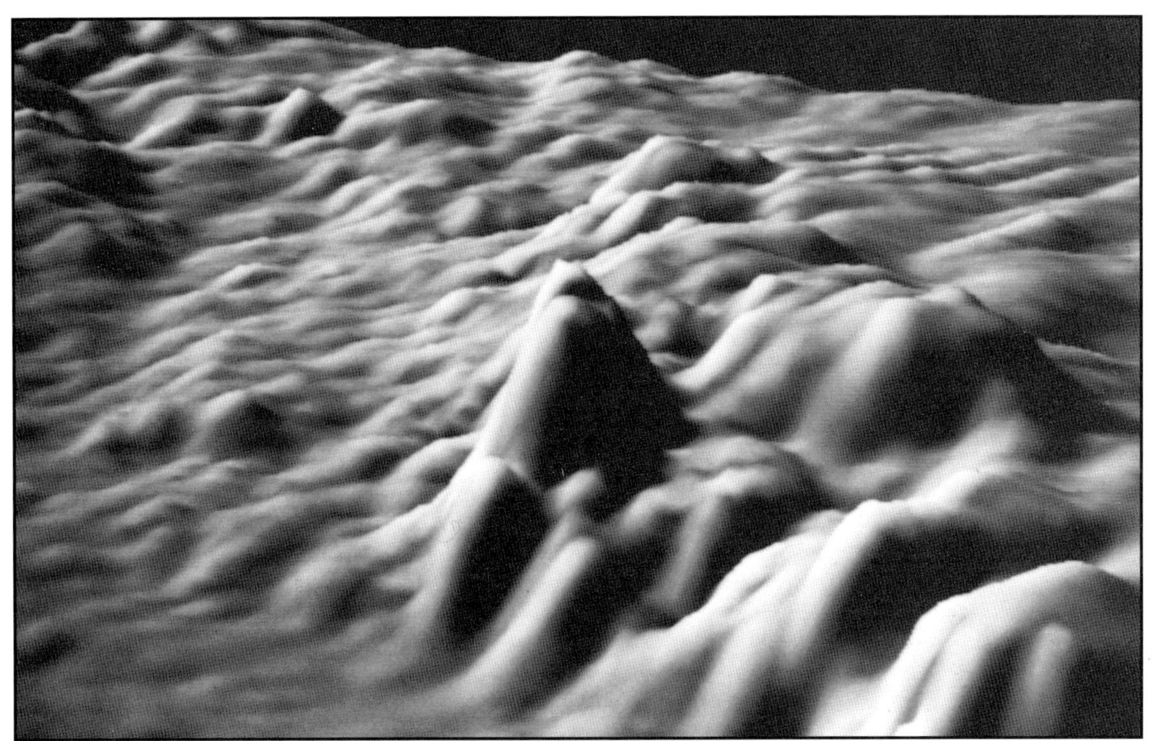

Hard Rocks

Finally, mention should be made of the importance of the rocky substrate to the biological community [Ricketts, et al., 1985]. Rocks present fundamentally different problems and opportunities to plants and animals that try to live on them than does unconsolidated material like sand and mud. Sessile (attached) organisms such as algae, stony corals, sponges, tunicates, hydroids, and even some mollusks require a firm point of attachment. In a very real way, we owe the present lavishness of the biota at Cordell Bank to the hardness of its ancient rocks, in turn determined by the deep and mysterious forces that generate such rocks in the cataclysmic collisions of continents.

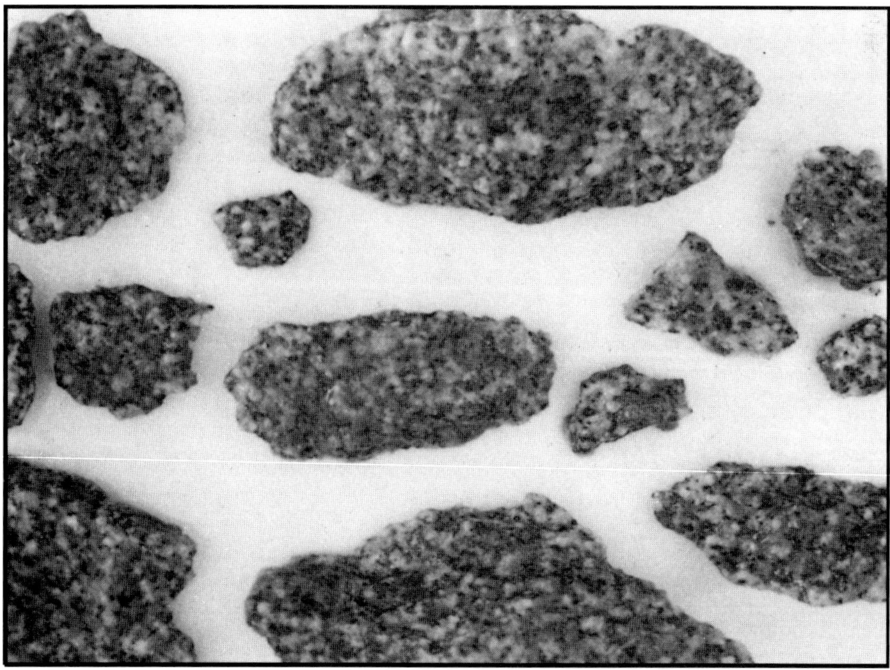

The bedrock from Cordell Bank. These granitic rocks are extremely hard. They contain plagioclase (30-40%), orthoclase (15-40%), quartz (25-30%), and biotite (5-10%). All the minerals show extensive deformation, which gives clues about the origin of the rocks. Besides granite, there is sandstone on the Bank, evidence of ancient erosion and recompaction of sediment.

RECENT HISTORY

Discovery

Cordell Bank lay unknown for millennia. Literally hundreds of mariners, possibly beginning with Francis Drake in 1579, passed directly over it without even suspecting its existence.

With the annexation of California into the United States in 1850, there was impetus to survey the coast to ensure maritime safety. The U. S. Coast Survey undertook the task. George Davidson was the most prominent of several men assigned to carry out the hydrography, and he was an observer without peer. On the night of October 20, 1853, as he was returning from a survey up the northern coast, Davidson found himself enveloped in a dense fog somewhere off Pt. Reyes. Dropping a lead line overboard in an attempt to determine his position, he found a mere 30 fathoms where he expected more than 60 (1 fathom = 6 feet, or slightly less than 2 m). He correctly realized that he had found a rocky bank, and he estimated its position (also correctly) as 37 km west of the Pt. Reyes head. He also correctly guessed (rather incredibly) that the minimum depth was about 20 fathoms.

Edward Cordell

No further attempt was made to locate the Bank until after the Civil War. In late 1868 new reports from mariners reached Davidson of a "shoal west of Pt. Reyes." By that time, Edward Cordell was engaged in extensive hydrographic surveys on the coast of California, reporting directly to Davidson.

Cordell had an interesting past. Born in 1828 in Philippsburg, Germany, he entered the famous Technisches Hochschule in Karlsruhe, but dropped out less than two years later. In 1849 he was indicted for high treason for his participation in the Revolution. Like many others in his position, he escaped to America. In 1851 he joined the U. S. Coast Survey. During 1854 he assisted Henry Stellwagen in the discovery of a bank (immediately named Stellwagen Bank) just outside Boston harbor. During the Civil War Cordell aided the Union Navy by surveying the harbors essential for replenishing supplies, sometimes under enemy fire. By 1865 he had so distinguished himself as an accomplished hydrographer that he was appointed an Assistant to the Superintendent and sent to California.

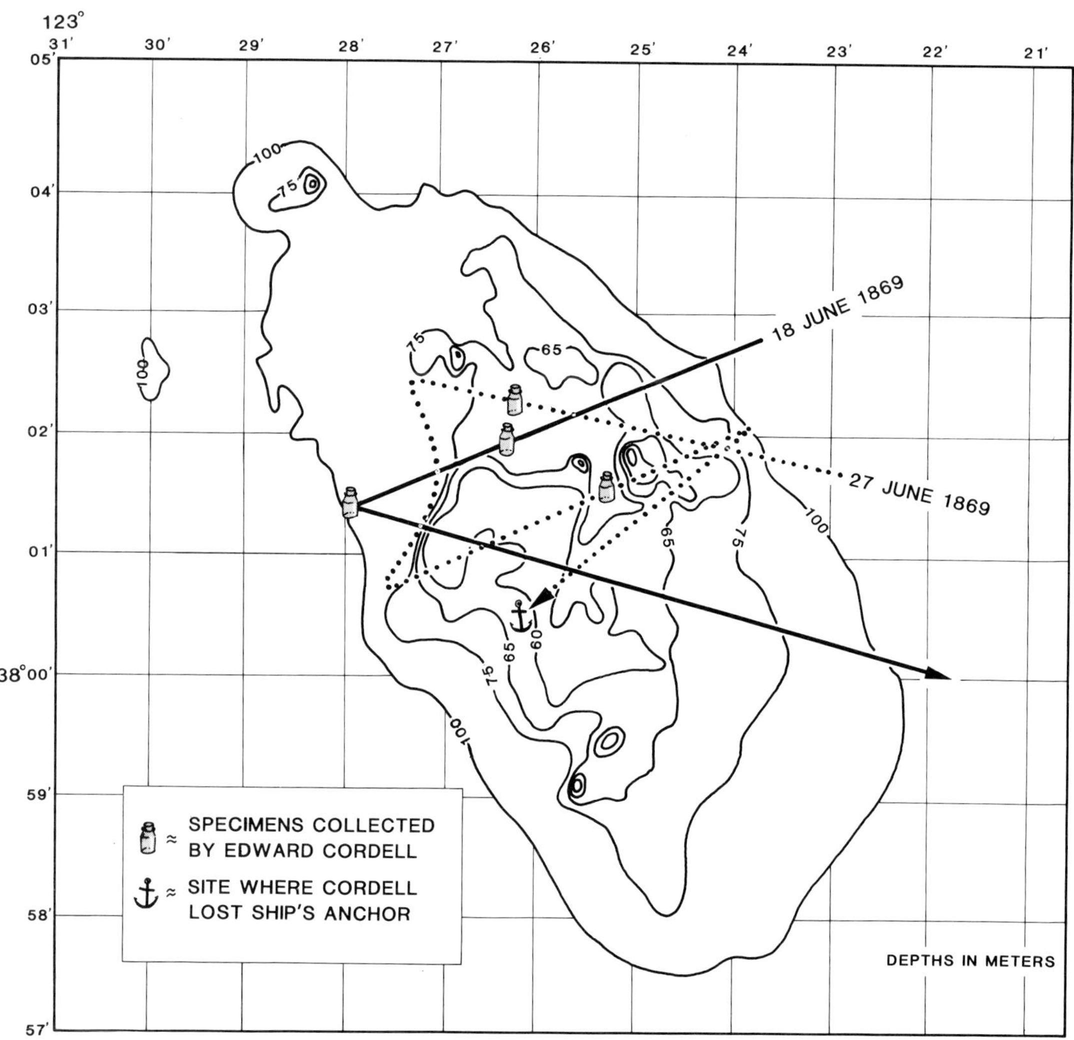

Events associated with the discovery of Cordell Bank by Edward Cordell in June, 1869. Cordell made two passes across the top of the Bank, stopping periodically to collect samples of the biota and put them in small jars. Four of these specimens are preserved in the National Museum of Natural History in Washington, D. C. On June 27, 1869, Cordell anchored overnight, but when a storm came up he was unable to recover his anchor and lost it when the line parted. Presumably the anchor is there today. The contours shown are those determined in the 1985 survey. The track of the vessel was taken from Cordell's log, positioned to best fit the contours determined in the 1985 survey.

Davidson suggested to Cordell that he search for the "shoal west of Pt. Reyes," and Cordell made several diligent searches, without success. Finally, during a two-week period in June, 1869, he succeeded in finding it and determining the overall contours. He was attracted to the location by the large numbers of birds and mammals, indicative of abundant food. Previous reports of discolored (green) water in the vicinity, often a sign of a biologically active spot, were in accord with his observations.

Edward Cordell died of an accidental fall in San Francisco just six months after rediscovering the Bank. Although Davidson thought the shoal should be named for himself, and an official in Washington thought it should be named Sutter Bank, Cordell's death obviated all these arguments. Within a year, Davidson and the others finally accepted the name that was already in general use: Cordell Bank.

Modern Expeditions

Study of the biology of Cordell Bank began with Cordell [Sources]. Using the sounding apparatus developed (and patented) by Stellwagen, he recovered eight samples of the biota, put them in vials, and shipped them off to the U. S. Coast Survey office in Washington. They were described as "red, slimy masses," and the Bank was described as "rocky, with live barnacles." Four of these vials still exist in the U. S. National Museum of Natural History in Washington, D. C.

Additional hydrographic surveys of Cordell Bank were made in 1873, 1911, 1929, 1960-62, and 1985 [Surveys]. No significant scientific research was done until 1949, when G Dallas Hanna of the California Academy of Sciences carried out a series of dredging expeditions in the Gulf of the Farallones to collect rocks [Chesterman, 1952; Hanna, 1952]. Incidental to the geological collections, a few invertebrates were obtained, and a few of these were added to the permanent Academy collections. No account of the biota was ever published.

The most extensive study of Cordell Bank was initiated in 1977 by the author [Schmieder, 1978, 1982, 1984, 1985, 1987, 1988]. Using air scuba, volunteers, and the optimism of inexperience, the first scientific dive on Cordell Bank was made on October 20, 1978. The sight we beheld that day was breathtaking: visibility greater than 25 m, jagged rocky outcrops with a dense and lavishly colored invertebrate and algal community, and a myriad of large rockfish circulating slowly overhead. A single glance was enough to convince even the most skeptical that this was truly an extraordinary place.

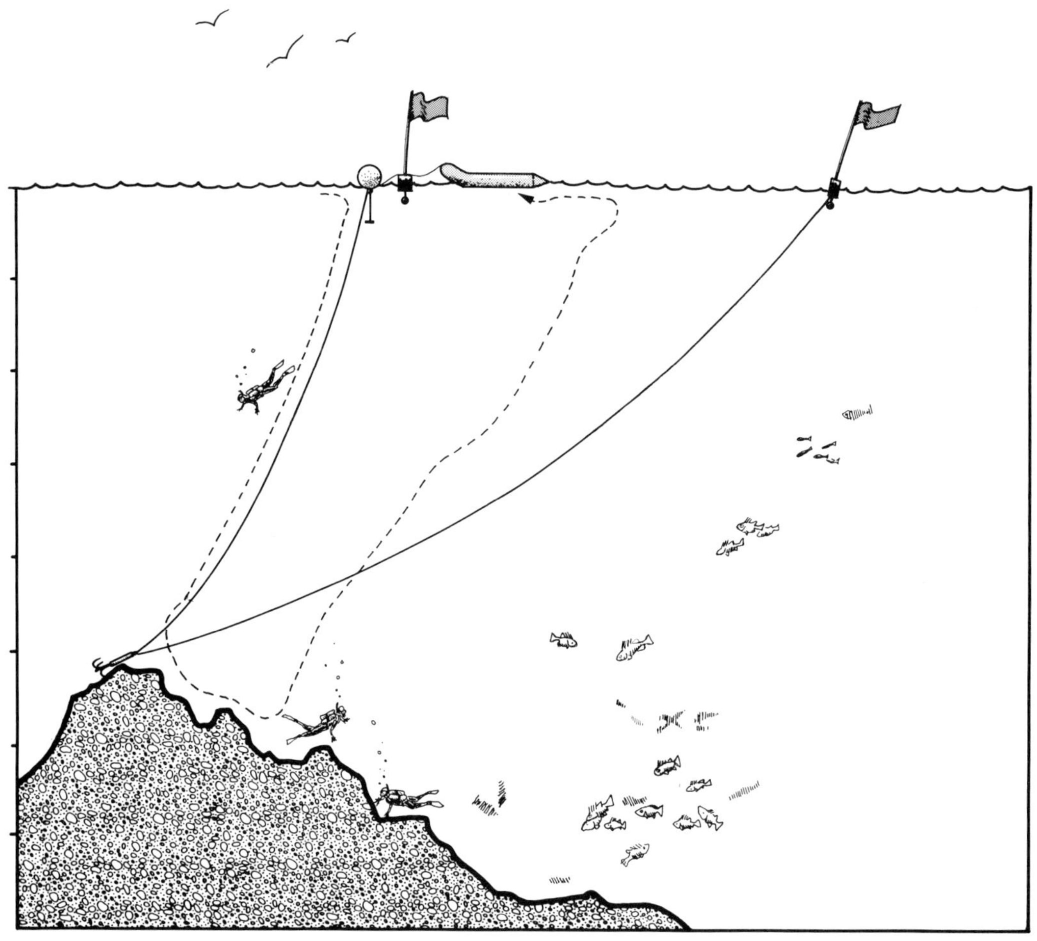

How divers explore an underwater island. The site is far from shore, tiny, and deep; it takes special techniques to locate the shallowest spots and establish lines for the divers. For safety, more than one line is used, and divers never venture far from the lines; getting lost in the open ocean would be very dangerous. A typical dive is to depths of 40-50 m, and lasts only 15 minutes. Divers must decompress before surfacing. The goal of each dive is to return with a bag full of specimens, a camera full of pictures, or a map of the topography.

Divers explore the top of a shallow ridge on Cordell Bank. Two of the divers are collecting representative samples of the plants and animals. Prominent in the view are wheel-shaped sponges, large-bladed algae, and many anemones. This scene was drawn from a verbal description before any bottom photographs were obtained. In retrospect, it is an amazingly accurate representation of the scene.

Beginning in 1978 the study of Cordell Bank was formalized under the protection of a non-profit association known as Cordell Expeditions, with the author as Director and Expedition Leader. In addition to diving to collect biological specimens and take photographs, we found it necessary to carry out high-resolution depth surveys to locate divable sites. The surveys revealed errors in the previous charts and a variety of previously unsuspected topographic features [Kruse and Schmieder, 1986]. Members of the California Marine Mammal Center participated in the expeditions by recording observations of marine mammals and seabirds [Webber and Cooper, 1983]. At the time of this writing, the species list for the Bank includes more than 450 known and new species of algae, invertebrates, fishes, birds, and mammals. A large collection of specimens has been distributed among various specialists. The biological community is documented in more than 3000 underwater photographs and considerable underwater film and video footage.

Cordell Bank was established as a National Marine Sanctuary on May 19, 1989, and provided with complete protection by an Act of Congress signed by the President on Aug. 9, 1990 [Federal Register, 1981, 1983, 1987, 1988, 1989, 1990].

What does it mean to be a "National Marine Sanctuary"? Interestingly, it means different things in different places. Each sanctuary has a unique set of regulations designed to address the unique needs of the site. In addition, the policy of the sanctuary program itself evolved over the course of designation of several sanctuaries, so the criteria and provisions varied. In the case of Cordell Bank, there is explicitly no restriction on fishing activities, or on normal traffic and commerce. There *is* prohibition against dumping damaging wastes, against taking plants and animals, against harassing birds and mammals, and damaging the fragile biological cover. There is also a prohibition against oil and gas activities. In fact, Cordell Bank is protected from oil industry activity by an Act of Congress, the only site within the jurisdiction of the United States so protected. Thus, in addition to the natural protection offered by its remoteness, there is legal protection of this area as well.

THE ENVIRONMENT

Subtidal

The most obvious aspect about Cordell Bank, of course, is that it's underwater. More than that, it's under *deep* water. This means that the community is isolated from the mechanical action of a crashing surf, and even from much disturbance due to wind-driven waves on the surface. Deeper than about half the wavelength of a surface wave, the motion of the water is very small. Thus, the hydrodynamic environment on the Bank is rather calm and nonviolent. The only significant flow is due to the relatively gentle, relatively large-scale turbulence generated by large-scale currents encountering a complex obstacle. Locally, on the scale of the benthic organisms, the flow is laminar—the organisms are free from the high stresses produced by waves, bubbles, high-velocity jets, tumbling rocks, and sediment scour.

Interestingly, divers regularly report feeling surge, even at depths up to 50 m. This mysterious phenomenon probably has two main causes: first, the turbulence associated with the large-scale flow. That is, as the water flows around the pinnacles and ridges, instabilities cause it to form swirling eddies on a scale of a few tens of meters; passage of the interface between two eddies would be felt by a diver as a surge. Second, very large swells *will* cause some surge on the bottom. It is relatively easy to estimate from linear wave theory that a swell of height 3 m and period 15 seconds produces a maximum horizontal velocity of the water of about 2 nautical miles per hour, or about 1 m/s. Traveling in water 100 m deep, this swell will produce horizontal particle velocities at a depth of 50 m of about 1/7 of the surface velocity, or about 0.3 nautical miles per hour, or about 15 cm/s, which will be quite perceptible to a diver.

Another important consequence of being deep is that the deep water is not completely saturated with oxygen. There is not the frothy foam that enables certain organisms to thrive in the splash zone, and there is not the energy associated with surf that can cause mixing, food distribution, and dispersal of propagules. The plants and animals in a deepwater environment are forced to find other ways to obtain food, carry on reproduction, and generally get around.

The environment at Cordell Bank is pacific. It is an attractive home for hordes of little, fragile creatures, and attracted to this quiet paradise, some big, husky ones as well.

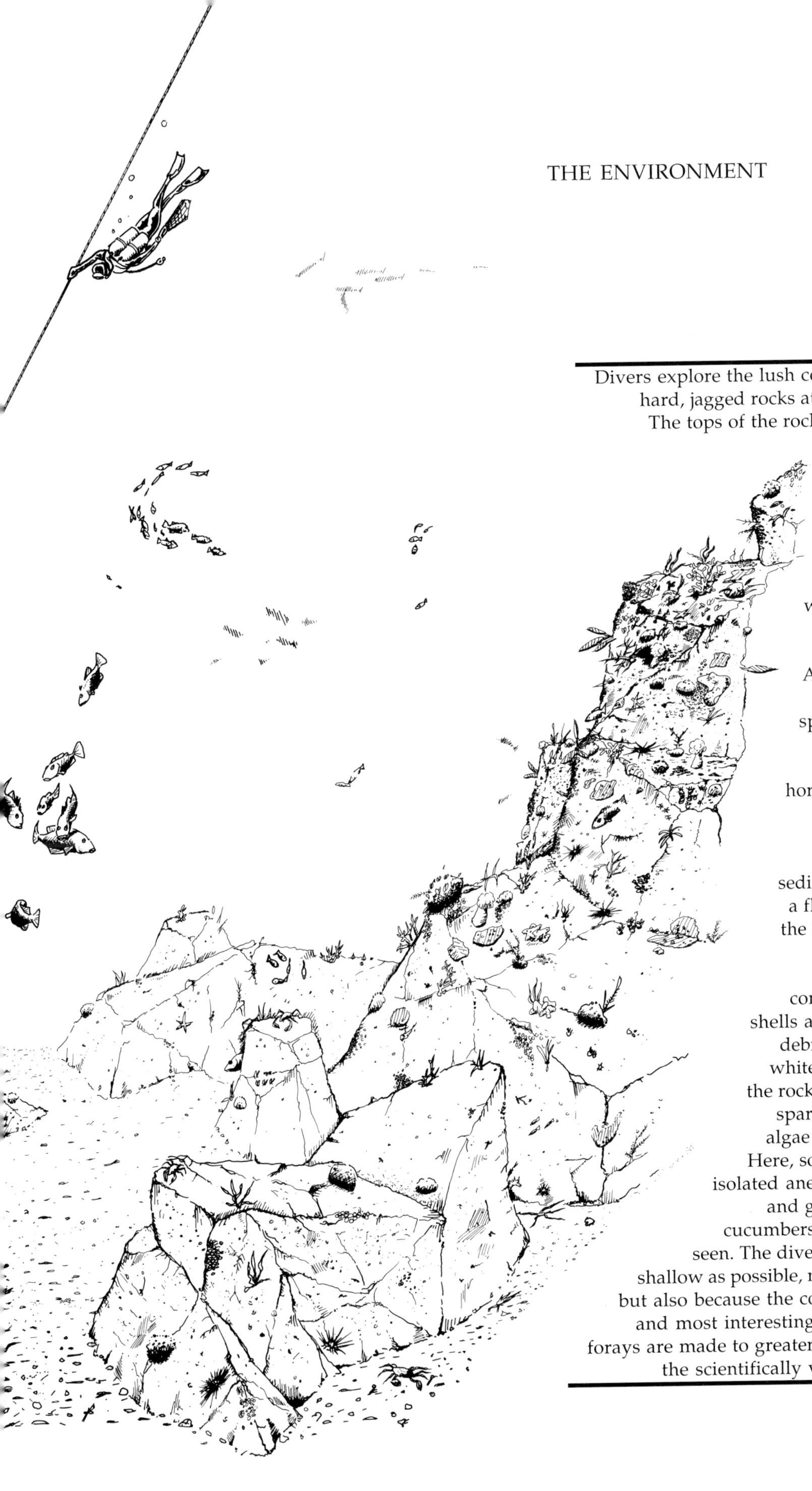

Divers explore the lush community on the hard, jagged rocks at a depth of 40 m. The tops of the rocks are completely covered with animals and a few plants. On the top of the shallowest pinnacle is a cap of barnacles, which successfully exclude almost everything else. A little deeper, the anemones and sponges dominate. The cover is very thick on the horizontal faces, and relatively thin on vertical faces. A deposit of coarse sediment fill provides a flat bottom around the bases of the rocks at 55 m depth. The sediment is composed of broken shells and other biogenic debris, and is brilliant white. Near the base of the rocks the cover is very sparse; red encrusting algae is the main cover. Here, solitary hard corals, isolated anemones, red algae, and grazers such as sea cucumbers and sea stars are seen. The divers try to remain as shallow as possible, not only for safety, but also because the cover is most dense and most interesting there. Occasional forays are made to greater depths to recover the scientifically valuable sediment.

Light

To a great extent (although not totally), the biological community now residing on Cordell Bank is determined by the present environment: the topography, substrate, water quality, light level, temperature, currents, etc. Therefore, it will be helpful in developing our understanding of the community to review what we know about the present environment.

If you were to dive on Cordell Bank, probably you would find three aspects most remarkable: the great visibility, the high light level, and the brilliant colors perceptible even at great depth. All these visual impressions have the same cause: the extraordinary clarity of the water.

The visibility in the water at Cordell Bank, at least during the late summer and fall, is generally greater than 20 m. It is more like the tropics than what we are used to in California. In the fall of 1983, the visibility was more than 40 m, although one year later it was unusually low—only about 15 m. Experienced divers know that the visibility in nearshore California waters is seldom more than 10 m [North, 1976]. The proximate cause of the extraordinary visibility on Cordell Bank is the strong current that is almost always present.

Visibility is determined by the concentration of suspended organisms and particulates—plankton, detritus, inorganic sediment, etc. These objects are generated locally on the Bank, and in oceanic and coastal water elsewhere and transported to the Bank by the currents. Regardless of their source, loose fragments, particles, and other debris are quickly carried away by the current. The current keeps the environment clean.

The high light level is no less remarkable. It is quite unnecessary to take a light when you dive, even to depths greater than 60 m. The visual experience at that depth is comparable to an early dawn on land. Again, it is the water clarity that is responsible. With very low particulate loading throughout the water column (and in the absence of a plankton layer near the surface), sunlight easily penetrates to bathe the biota in a diffused illumination anomalously high for California. The fact that this is the normal state of affairs on Cordell Bank is shown by the presence of numerous organisms at significantly greater depths than they are normally found elsewhere.

Color

But by far the most remarkable visual experience is seeing the blaze of color on the bottom, even at depths greater than 50 m—purple, blue, green, yellow, white, and red. No diver returning from his or her first experience at Cordell Bank fails to be awestruck by the color. The origin of these colors was a mystery for several years since it is common knowledge (at least among divers!) that only blue and green light penetrate more than a few m of seawater. Why should one see such vivid red color so deep?

Many people thought that nitrogen narcosis, a condition that affects almost every diver breathing air at depths greater than 40 m, was at least partially responsible. One of the manifestations of narcosis is an intensification of color perception: reds seem redder, whites seem whiter, and so on [Dueker, 1970]. Another factor may be the divers' knowledge or expectation of what colors the organisms are supposed to be.

Eventually it became clear that the color was a real phenomenon: the divers really were seeing colored light. Part of the explanation is that purple colors can be generated easily from the penetrating blue and violet light. Green and yellow light, while reduced in intensity, are still sufficiently strong to be perceptible to a dark-adapted eye. The white appearance of some objects is often generated by structural features such as microcrystals [Nicol, 1960]; those organisms are white for the same reason that snow is white.

But none of these explanations can account for the observed brilliant red colors. Red light is rapidly filtered out by seawater, even clean seawater. In normal oceanic water such as that at Cordell Bank, less than 1/1000 of the red light (wavelength 650 nanometers) penetrates to 50 m. Without the red light, how could we see objects on the bottom as red? After years of pondering, we finally found the solution to the mystery: They appear red because they create the red light themselves! A more detailed explanation will be given in the next chapter. For now, it is enough to say that because of the various factors producing the variety of colors, the landscape on the bottom appears to divers to be bathed in broad daylight. There is little sensation of being under 50 m of ocean water.

Living on the Edge

Living on Cordell Bank is a little like living in Switzerland: a large part of the real estate is vertical. True, overall the Bank is relatively flat, the result of erosion during hundreds of thousands of years. But the top of the Bank is actually very rough, cut by deep gullies, channels, holes, and cliffs. Most of the animals larger than a few cm, including the sponges, hydrocoral, tunicates, hydroids, crabs, and mollusks, clamber for space on the tips of thin ridges and tiny pinnacles, some of which reach to about 40 m. By far the great majority of species and biomass is found shallower than 50 m. The cover is nearly homogeneous above 45 m.

The shallowest points on the Bank lie along a central ridge or escarpment that runs the entire north/south length of the Bank. Associated with this feature is a series of faults and other erosional structures.

Close-up photograph of a sample of the sediment from Cordell Bank. The material is composed of broken shell fragments and other pieces of calcium carbonate. The sediment deposits are especially interesting because they contain the integrated fall-out from the ridges above them over many thousands of years. Thus, their stratified layers should contain valuable information about the variation of the biotic populations on the ridges with time.

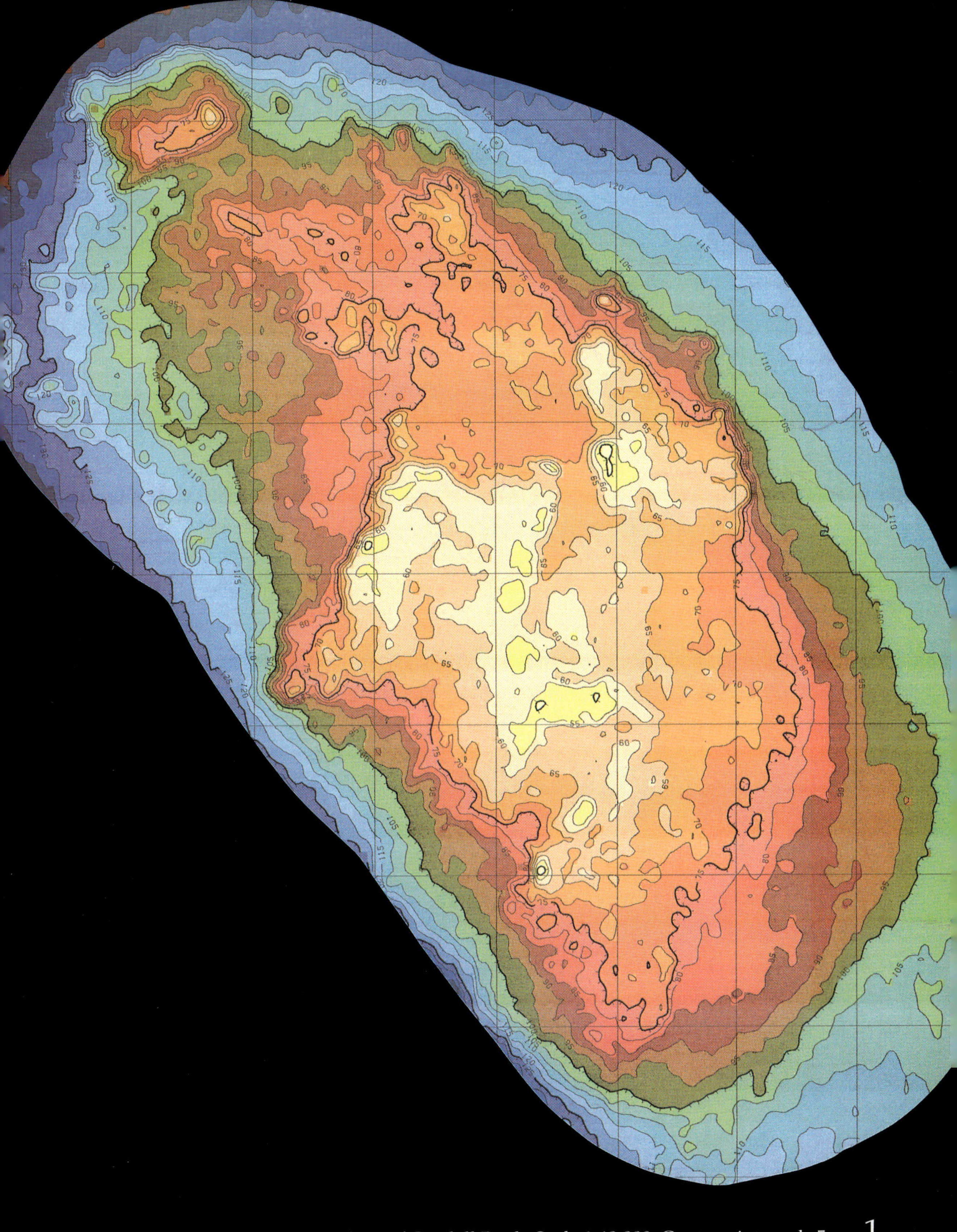

1. Bathymetric chart of Cordell Bank. Scale 1:40,000. Contour intervals 5 m.

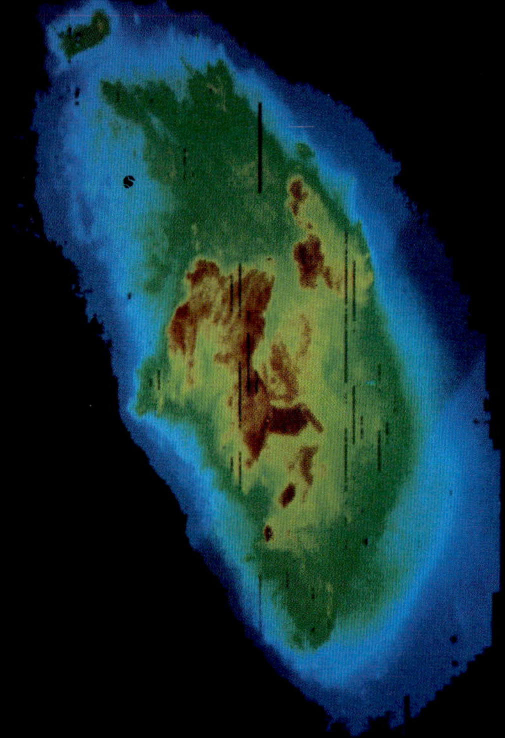

2. Areas accessible with SCUBA (shown in red).

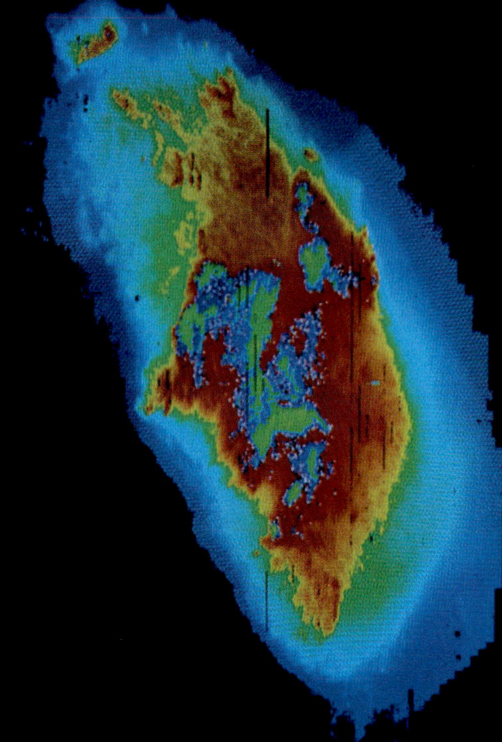

3. Areas accessible with ROV (shown in red).

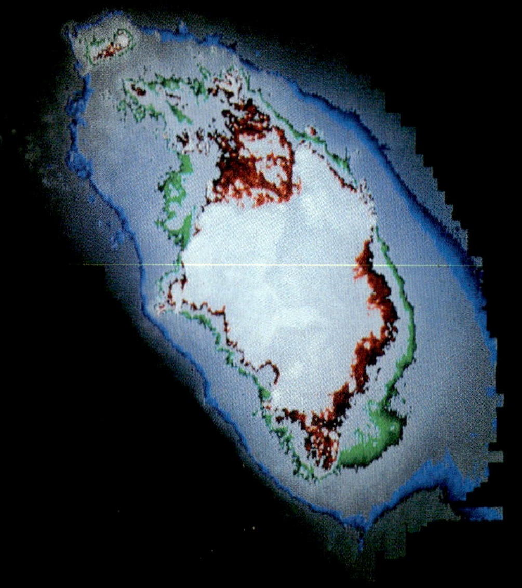

Computer images of the bathymetry of Cordell Bank. Images were constructed using a minicomputer and a data base of 6 million individual soundings. In these images the depths are color-coded. The horizontal resolution of these plots is about 4 m.

4. Selected contours highlighted to aid in study of wave-cut terraces.

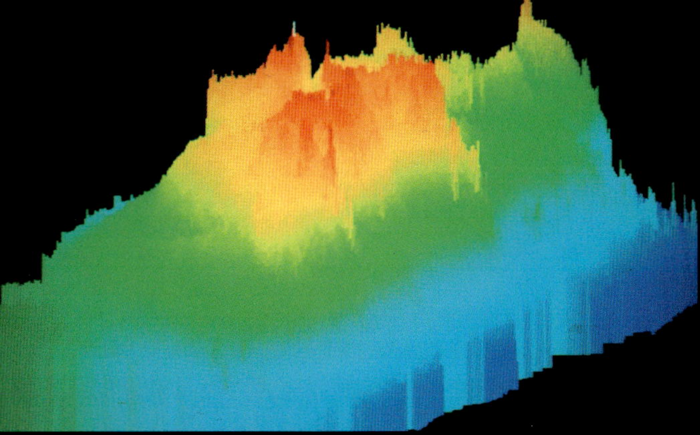

5.

Entire Bank topography.

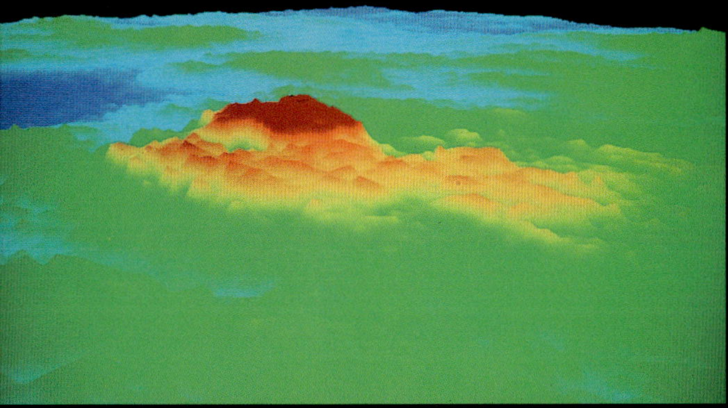

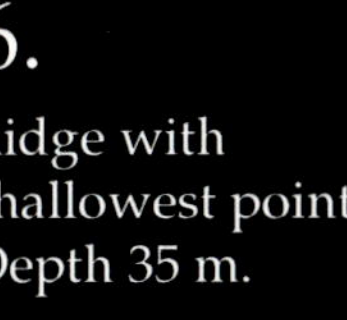

6.

Ridge with shallowest point Depth 35 m.

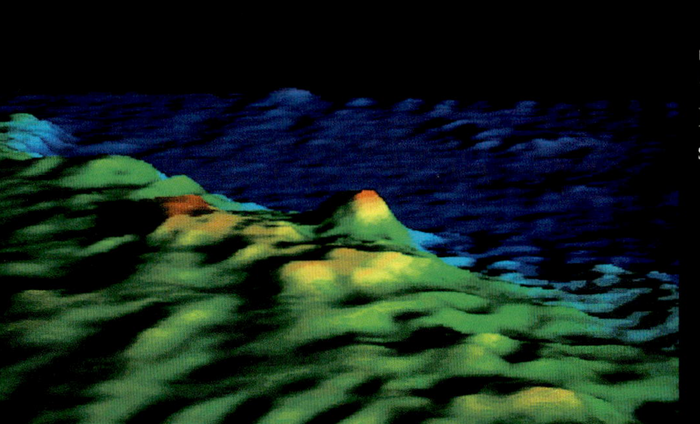

7.

Pinnacle on west side of Bank.

8.

Drainage channel

9.

Pinnacle named *Tor Hakluyt* by the author.

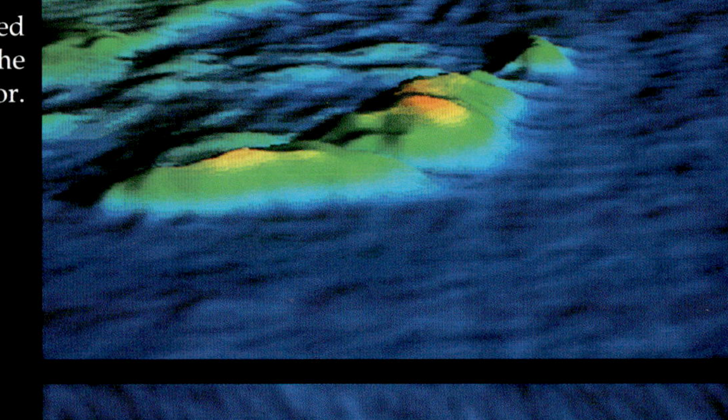

10.

Top view of *Tor Hakluyt*.

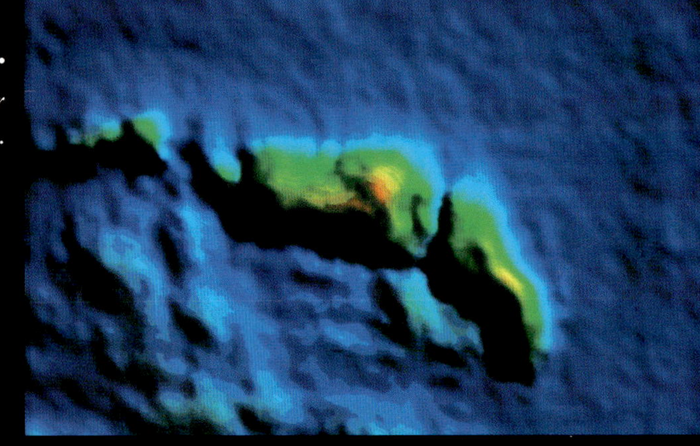

11.

Pinnacle near central plateau.

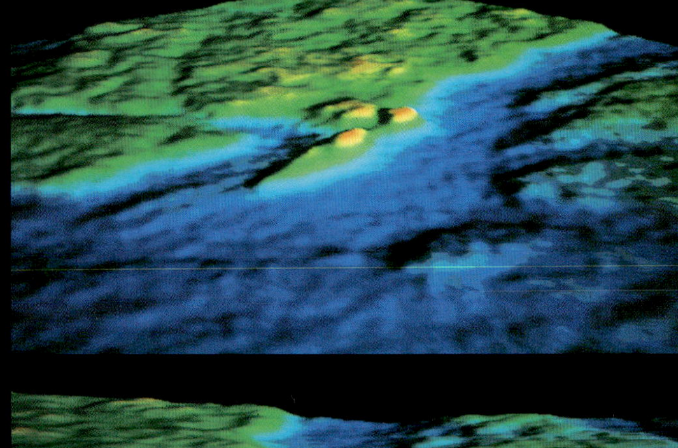

12.

Another view of central plateau pinnacles.

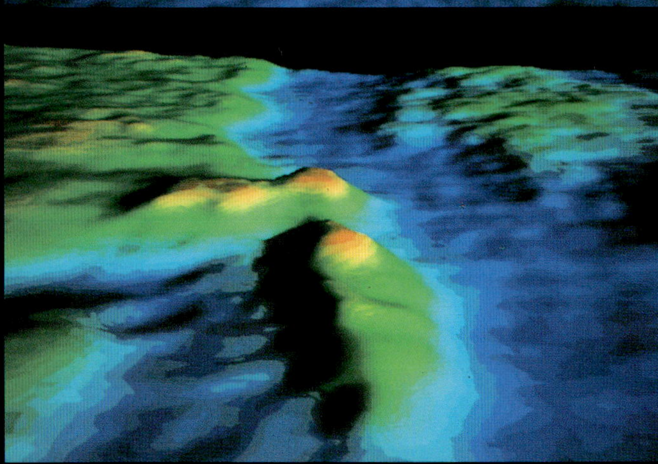

Photograph of the waters over Cordell Bank, with an artist's 13
representation of the underwater island below.

Unidentified coronate 14
syphomedusan (jellyfish).

15. Pacific white-sided dolphin.

16. Measuring the distances between peaks. Depth 40 m.

17. Encountering animals on a shallow pinnacle.

Passing a sponge and anemone covered ridge. 18.

Studying the sponges, anemones, and hydroids. 19.

20. Narrow sediment-filled chasm. Bottom at 55 m.

21. Reaching for a remote ridge. Sediment at 55 m.

22. Sampling the debris in a large hole. Depth 42 m.

Encountering a small sediment pocket. Depth 40 m. 

Craggy rocks, a remnant of surf erosion. Depth 40 m. **24.**

Sampling the biota. Depth 38 m. **25.**

26. Diatom *Entopyla* and red algae.

27. Brown alga *Desmarestia tabacoides*.

28. Red alga *Maripelta rotata*.

Yellow sponges, with *Corynactis californica*. 29.

White sponge *Stelleta clarella*. 30.

Globular brown and yellow sponges. 31.

32. Aggregate: sponges and a white tunicate.

33. Aggregate: hydrocoral, sponges, anemones.

34. Aggregate: hydrocoral, sponges.

Aggregate: sponges, hydrocoral. 35.

Aggregate: sponges, hydroids, hydrocoral. 36.

Aggregate: sponges, anemones. 37.

38. White-plumed anemone *Metridium senile* and the red sea star *Mediaster aequalis*. The star is known to prey on the anemone. Several colormorphs of the California hydrocoral *Allopora californica* are seen. Depth 50 m.

Globular clonal aggregate of the anemone *Corynactis californica*. **39.**

Anemone *Corynactis*, showing oral opening. **40.**

Corynactis, fully extended. **41.**

42. *Corynactis*. Strobe light. Depth 40 m.

43. *Corynactis*. Available light. Depth 40 m.

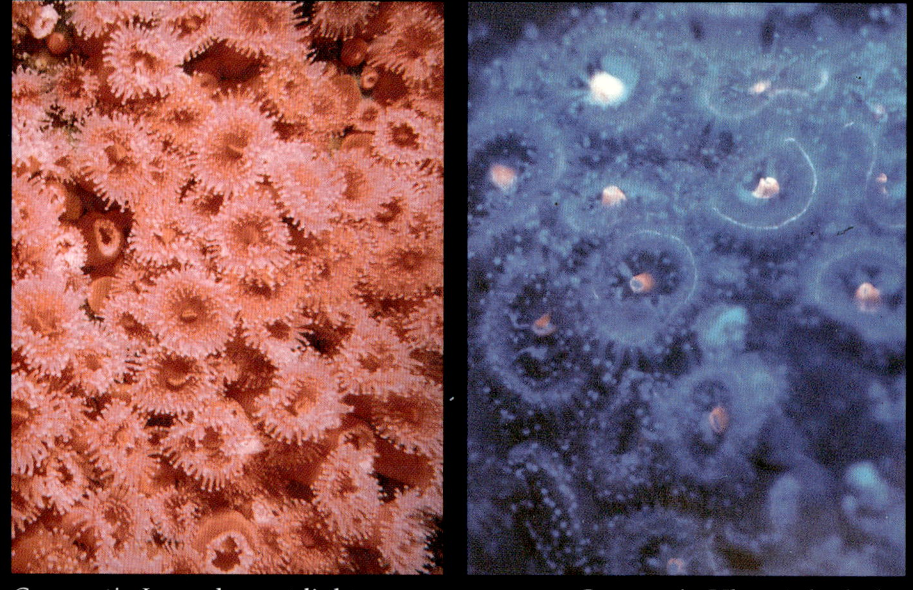

44. *Corynactis*. Incandescent light.

Corynactis. Ultraviolet light. 45.

Snail *Pedicularia californica,* in *Allopora* habitat. **46.**

Top shell *Calliostoma annulatum.* **47.**

48. *Calliostoma ligatum.*

49. Field of hydrocoral *Allopora californica*. Depth 40 m.

50. California hydrocoral *Allopora californica*, with hydroid *Garveia*.

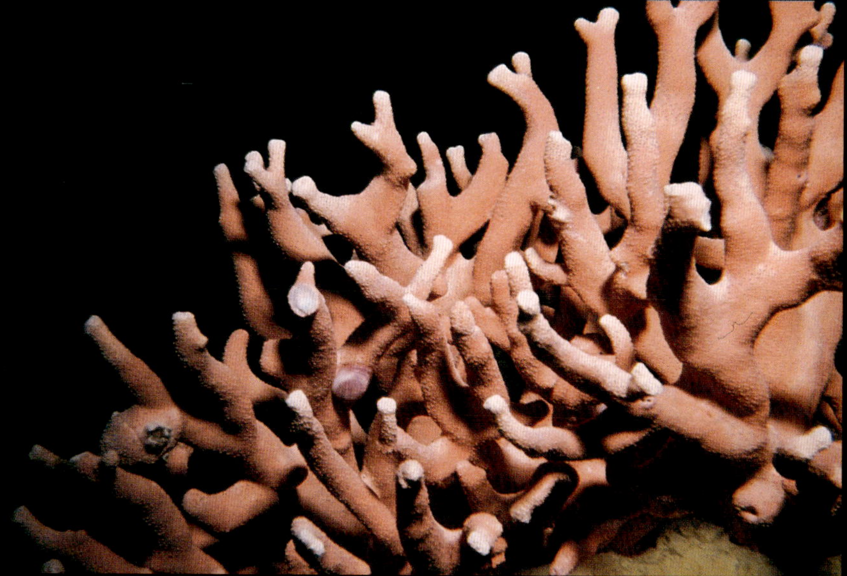

51. *Allopora californica*, with *Pedicularia*.

Yellow hydroid *Garveia annulata*. Depth 40 m. 52.

Hydroid *Garveia annulata*, with *Corynactis*. 53.

Hydroid *Garveia annulata*. 54.

55. Urchin *Strongylocentrotus* attacking sponge.

56. Close-up of previous picture.

57. Area being kept clear by chiton *Tonicella*.

58.

Decorator crab *Loxorhynchus crispatus*.

59.

Decorator crab *Loxorhynchus crispatus*.

60.

Tunicate *Cystodytes lobata* on hydrocoral.

61.

Tunicate *Cystodytes lobata* on hydrocoral.

62. Rockfish. Available light. Depth 40 m.

63. Rockfish. Available light. Depth 40 m.

64. Rockfish. Available light. Depth 40 m.

Yellowtail rockfish *Sebastes flavidus*. Strobe light. Depth 40 m. 65.

Juvenile rockfish. Strobe light. Depth 40 m. 66.

Lingcod and Yellowtail rockfish. Strobe light. Depth 40 m. 67.

68. The author on the shallowest point on Cordell Bank. Depth 36 m.

69. Barnacle *Balanus*.

70. Barnacle *Balanus*.

Widow rockfish. A double exposure captures the sense of transition without alteration: 71.
millions of individuals have come and gone, but the scene is indistinguishable from that eons ago. Some of the many mysteries of underwater islands are thereby presented to us: Is the community in equilibrium, or is it changing at a rate just beyond our ability to measure it? If it is changing, what is the nature of the change, and what is causing it?

72. Installing a fish trap. Depth 45 m.

73. Rosy rockfish *Sebastes rosaceus* captured.

74. Rosy rockfish *Sebastes rosaceus*.

Painted greenling *Oxylebius pictus.* 75.

Pacific electric ray *Torpedo californica.* 76.

Yelloweye rockfish *Sebastes ruberrimus.* 77.

78. Sediment deposit. Depth 58 m.

79. Scallop in sediment.

80. Coarse sediment.

81. Sediment trap.

82. Brachiopod in sediment.

83. Fine sediment.

Anchor lost on pinnacle. Depth 40 m. 84.

Chain lost with anchor. Depth 38 m. 85.

Fishing weight with line. Depth 45 m. 86.

87. Taking samples in hole. Depth 42 m.

88. Diver takes self-portrait standing in hole.

Hole. Depth 43 m. 89.

Hole. Depth 40 m. 92.

Hole. Depth 42 m. 90.

Hole. Depth 43 m. 93.

Hole. Depth 41 m. 91.

Hole. Depth 43 m. 94.

95.

General scene on the Pt. Sur bank. Depth 38 m.

96.

Hydrocoral *Allopora californica*. Depth 36 m.

97.

Spectacular topography on the Pt Sur bank. Available light, 1/60 s exposure. Depth 45 m.

98.

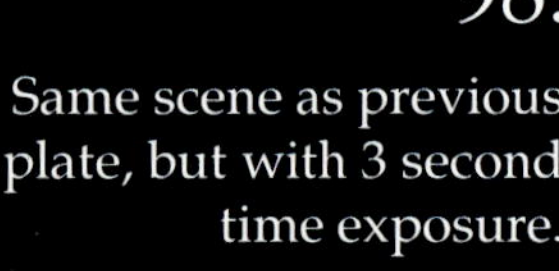

Same scene as previous plate, but with 3 second time exposure.

Sediment

The sides of many of the ridges on the Bank are extremely steep: on the shallowest ridge at the north end they drop on both sides at angles greater than 70°. Just below Craine's Point at the south end is a cliff that is precisely vertical for more than 40 m. Collected in the spaces between steep walls are pockets or channels of sediment. The contact between the sediment and the wall is always very sharp; the sediment fills the basins like a white liquid.

The sediment gives evidence of strong currents; even at depths greater than 60 m, the sediment is piled into low ripples or dunes. The ripples are typically 10 cm high and spaced 1 m apart. From measurements of size and density of the grains, it is possible to calculate the speed of the bottom current that built them [Heezen and Hollister, 1964]. The result in one case: about 1 knot (=1 nautical mile per hour, about 0.5 m/s).

Cordell Bank is relatively unaffected by the massive outpouring of sediment from San Francisco Bay [Carlson and McCulloch, 1974]. The plume emitted from the Golden Gate extends about halfway to the Farallon Islands, but does not reach far enough west or north to affect Cordell Bank. Furthermore, although turbid water is generated all along the northern California coast, it seldom gets more than a few km offshore [Carlson and Harden, 1974]. Cordell Bank is just too far offshore to get dirty.

There is one exception to the previous statement. Satellite photos sometimes show a marked plume of turbid water emanating from Pt. Arena, moving south along the coast, and then veering westward, directly over Cordell Bank. During such times, it would be expected that the sediment load at the Bank would be well above normal, and the visibility would be sub-normal.

A sample of the hard corals found in the sediment deposits. This coral, *Balanophyllia elegans*, forms individual cups of calcium carbonate in which the animal lives. When the animal dies, the cup remains attached to the substrate, but may be dislodged by a fish, crab, or perhaps another growing organism. The cup then falls or rolls downslope until it reaches the flat sediment deposits, where it remains relatively intact. Often tiny single-celled animals called foraminifera (see page 61) will populate the protective recesses of these skeletons; when the coral is collected and dried, the foraminifera can be shaken out like salt.

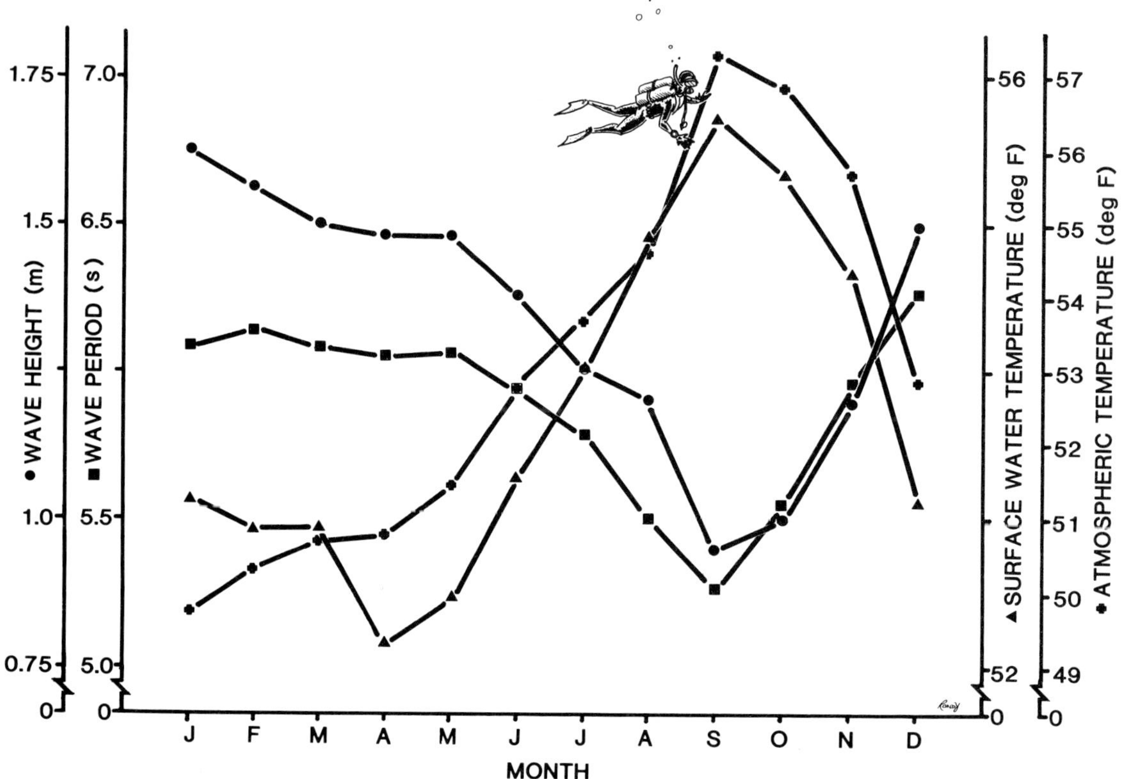

Temperature

The temperature of the water is almost always 52-56°F on the surface [Galloway, 1977], dropping perhaps 5°F toward the bottom. There is sometimes a thermocline (a relatively sudden change in temperature with depth); divers have sometimes reported two thermoclines.

There is a rather spectacular exception to this norm: El Niño. Every few years (four, on the average), the equatorial east winds are anomalously low, and water that is normally driven westward across the Pacific backs up against the west coasts of the Americas. The biological and meteorological consequences of the 1982-83 event were heroic [Barber and Chavez, 1983; Cane, 1983; Rasmussen and Wallace, 1983; Feldman, Clark, and Halpern, 1984]. In October 1983, the surface water temperature at Cordell Bank was above 65°F, almost 10°F higher than normal. Enjoying all this warmth were more sharks than really seemed necessary!

Variation of average environmental conditions near Cordell Bank through the year. The months are indicated along the bottom. The left vertical scale refers to the surface waves, which are seen to be calmest in September. The right vertical scale refers to the temperatures, which are maximum in September. These data show clearly why divers prefer to work during the fall, but they also indicate that there probably is a variation of the biota through the year. Therefore, populations in September may not be precisely the same as those in other months.

Currents

The animals on Cordell Bank have to hang on for dear life, as currents on and around the Bank are usually quite strong (up to 2 knots). Much of the year the current moves from north to south, and is referred to as the *California Current* [Jennings and Schwartzlose, 1960]. This current is the eastern part of the huge wind-driven circulation system of the North Pacific. The water is pushed eastward toward Alaska, turns south down the coast toward Oregon and California, then converges with the westward-moving Equatorial Current to move back across the Pacific. On the California coast, this current extends out to about 800 km.

It is not quite that simple, of course. It has long been known that when a current flows parallel to a coastline, the Coriolis force (due to the rotation of the Earth) deflects it to the right (in the northern hemisphere). On the California coast, this means that surface water driven south by the wind is pushed west, i.e., offshore. This water is replaced by deep offshore water forced to move up toward the shore by the pressure of the excess surface water above it. This process is known as *upwelling,* and it is a very important process in the maintenance of biological communities on the continental shelf [Suess and Thiede, 1983]. Off the central California coast, upwelling usually occurs from March through August.

But it is even more complicated than that. In addition to upwelling, the coastal current generates a counter current, moving opposite to the surface current [Reid, 1962]. In California it is known as the *Deep Countercurrent* or the *California Undercurrent*. It is about 80 km wide and moves northward at up to 1 knot. It is believed that when the winds die down in late fall, the Deep Countercurrent rises to the surface, producing the reverse flow known as the *Davidson Current* [Reid and Schwartzlose, 1962]. This phenomenon lasts until early spring (February to April).

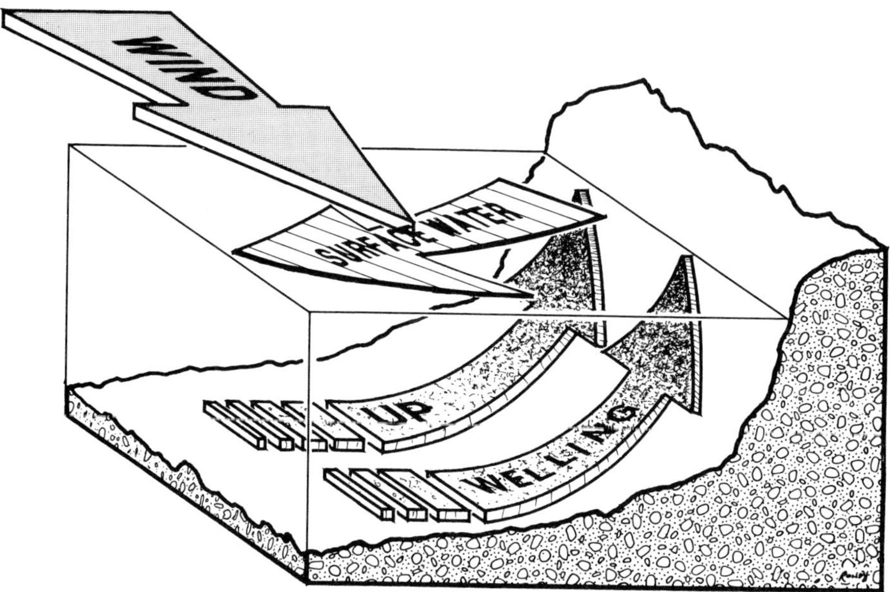

The process of upwelling off the northern California coast. The wind and Coriolis force drives the surface water offshore, pulling deep water in toward shore. The deep water is extremely cold, and carries with it dissolved organic materials that can be utilized by plants and lower animals to synthesize tissues. The water thus acts like a liquid fertilizer and contributes significantly to the support of the biota on the continental shelf.

As if this were not complicated enough, recent satellite infrared imaging has shown that the California Current is not a simple uniform southward flow. It actually consists of "intense meandering current filaments intermingled with synoptic mesoscale eddies" [Mooers and Robinson, 1984]. This means that the surface waters off the coast form huge turbulent eddies similar to those seen in smoke columns, boat wakes, and the Voyager photographs of Jupiter. These eddies rotate, interact, and evolve as they drift. Squeezed between them are narrow regions of cold upwelled water that are stretched into long sinuous "jets" often hundreds of km long. Cordell Bank is a sufficiently large obstacle to modify these eddies and jets significantly. In fact, under proper conditions, obstacles in flowing fluids generate vortices. Cordell Bank, like other banks and islands of California, could be a generator, as well as a receptor, of turbulent vortices and jets.

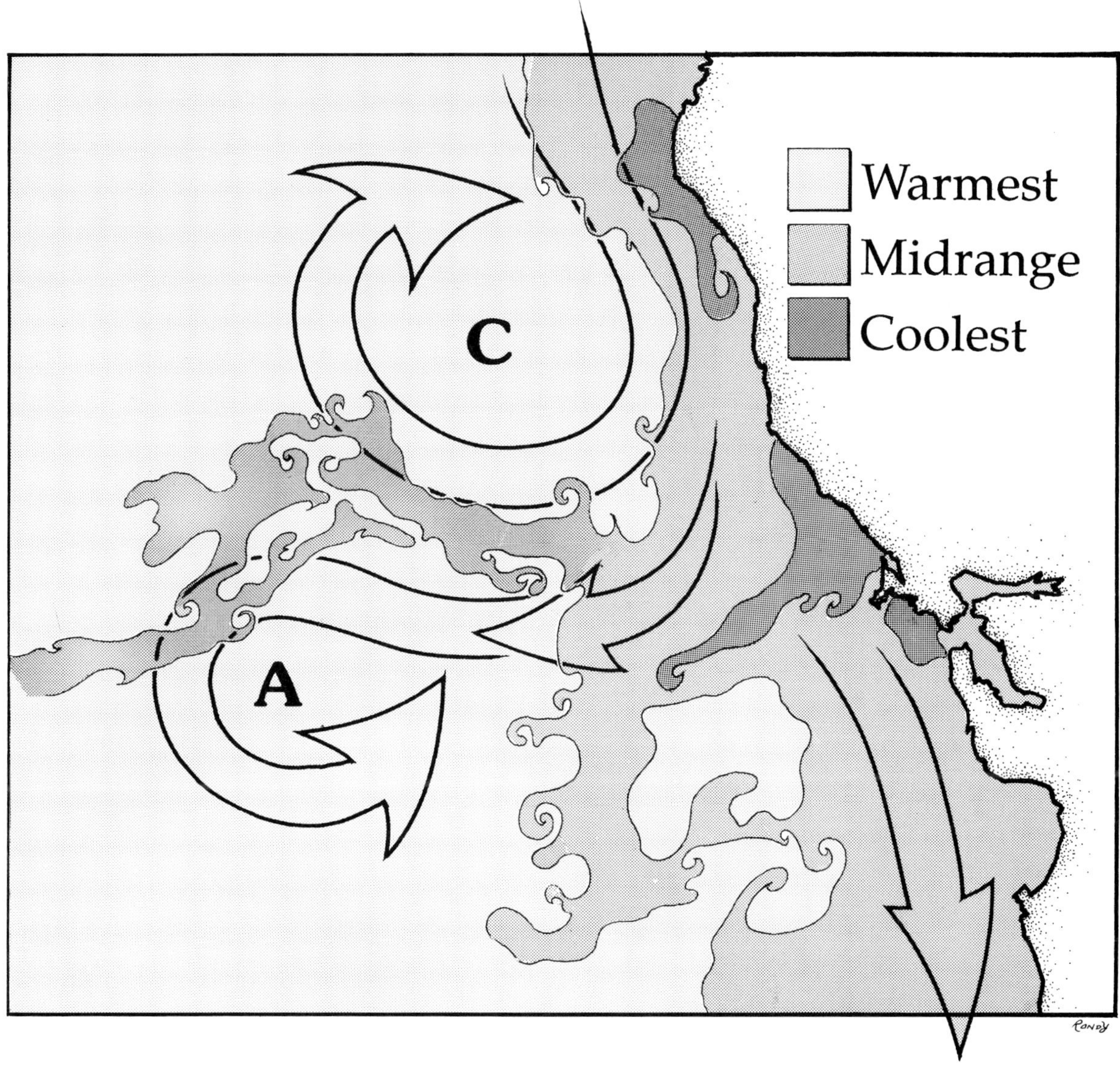

The flow of oceanic water off the northern California coast. A typical pattern is shown, this one seen in August, 1982. These patterns evolve in time as the eddies and jets swirl in turbulent flow. The letter "C" represents a cyclonic eddy; "A" shows an anticylonic eddy. Between them is a mesoscale jet of slightly warmer water. The jet originates from flow along the coast near Pt. Arena, which swings out across Cordell Bank.

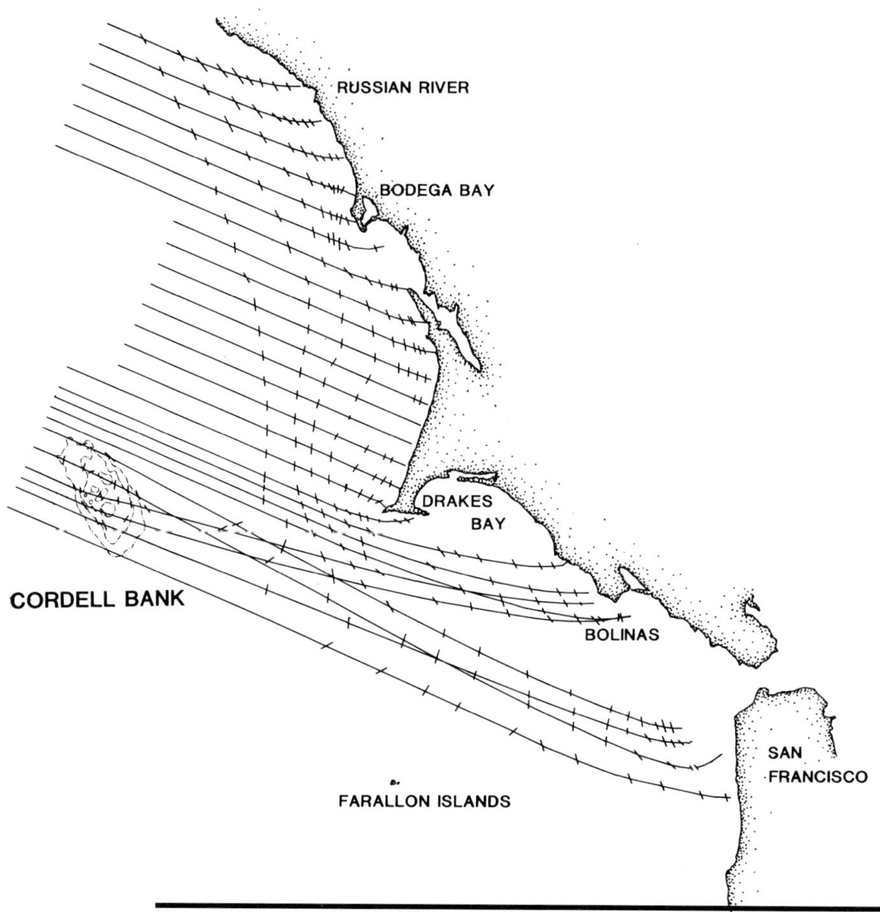

Refraction of the surface waves as they pass over Cordell Bank. The paths of the advancing wave crests are shown as lines; the wave crests as short cross lines. As the waves advance from the open ocean, they encounter the shallower water on the Bank. These waves slow, while those passing to the sides proceed unimpeded. The net effect is that the Bank acts like a lens, bringing the waves to a focus. The location of the focus will, of course, depend on the wave speed and direction.

The refraction of waves by the Bank is clearly seen in aerial photographs: the Bank acts like a lens and focuses the waves as they proceed over and past the Bank [Johnson, 1953]. On a much smaller scale, individual ridges and pinnacles produce refractive effects in the water flow: vessels trying to run over a high point find themselves pushed off to the left or right, then pulled back on the far side. These refractive flows seem to generate a "force" which mystified many early sailors who were unaware of the existence of the underwater mountain.

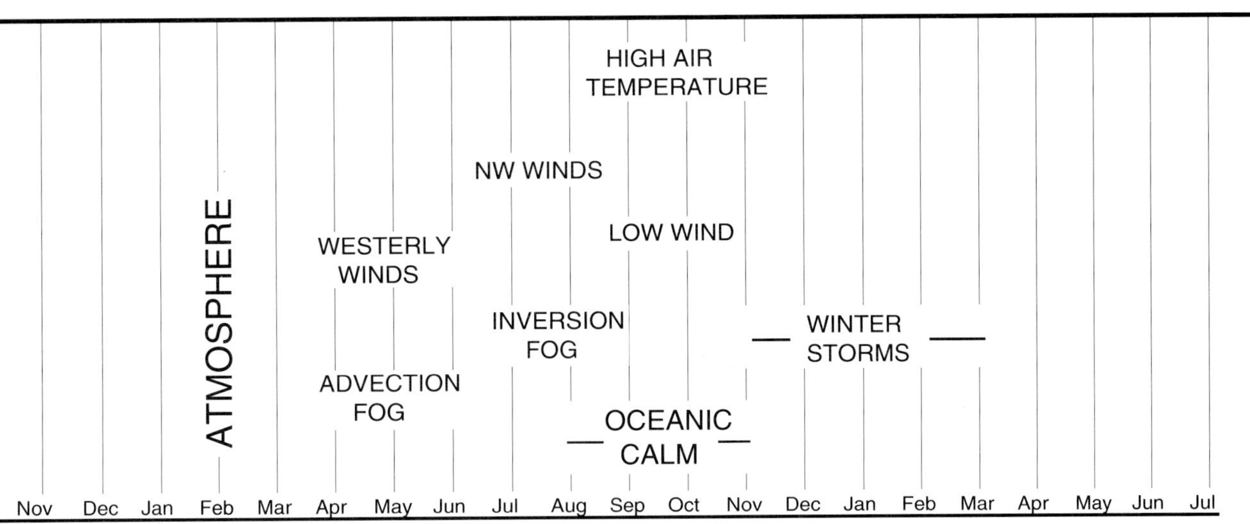

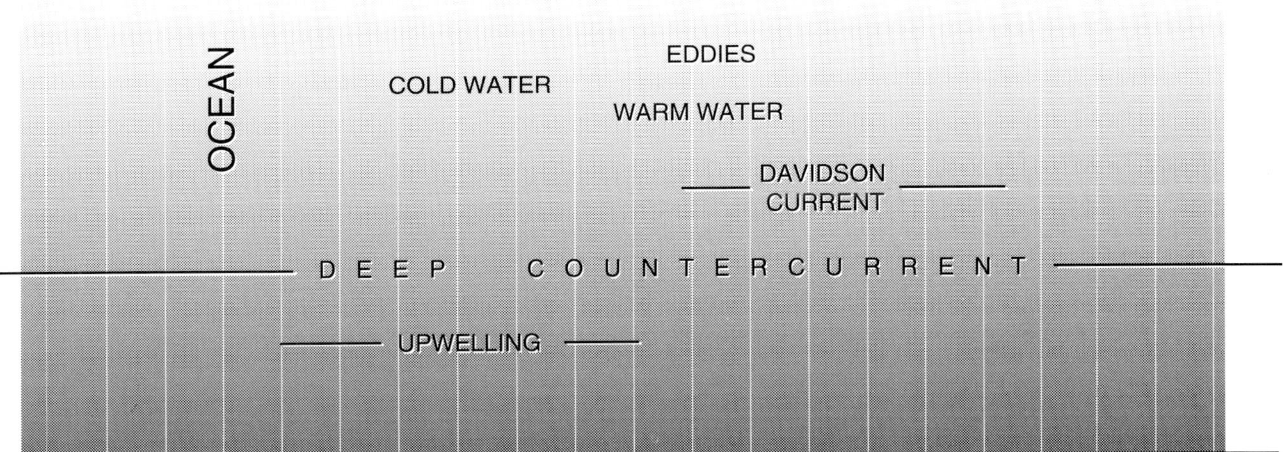

Variation of some environmental conditions through the year. The months are indicated along the center line; above this line are atmospheric factors, below are oceanic factors. This diagram shows, as does the figure on page 38, that the period September-October is the calmest and warmest. The transition to the winter storms usually is sufficiently sharp that news media remark on "turning the corner."

Going There

Perhaps the most appropriate description of the currents on and around Cordell Bank is "highly variable, unpredictable, and undependable."

Fortunately for Cordell Bank explorers, there is a respite from this hydrologic turbulence. In late September and October the winds calm, and the surface current and waves die down. This period is sometimes called the *Oceanic Period*. The Deep Countercurrent begins to rise, providing a short quiescent period when the current at the surface is practically zero. The water temperature is the highest of the year [Galloway, 1977], due to cessation of upwelling. Many oceanic animals begin migrations. The ocean can be so smooth that you can see the ripples spread from seabirds bobbing on the surface. This is the best time of the year to go to Cordell Bank. But do not delay; once the corner is turned into the winter weather pattern, it is one storm after another.

The easiest, and generally the only, way to get to Cordell Bank. Here, the author's vessel, the *Cordell Explorer*, is in the midst of an active expedition off Pt. Reyes.

ENVIRONMENTAL EFFECTS ON THE BIOTA

Community

Any community is the product of numerous evolutionary processes: competing species undergo genetic alteration to form new species; noncompeting species form commensal and host-parasite associations; predator-prey relationships shift in response to changing community composition; immigration introduces species new to the community and extinction removes other species [Moore, 1958]. These processes occur at rates that also depend on the external pool of organisms, which itself is changing. Finally, operating on the pool of organisms made available by these evolutionary processes, the present local environment determines the structure and function of the community.

Considering Cordell Bank as a very small ecosystem in communication with a large and stable reservoir (the entire continental shelf and the rest of the ocean), we infer that the exact community composition would be much more sensitive to small changes in the environment than would a larger ecosystem [Cox and Moore, 1980]. A small shift in sea level, for instance, would have a major impact on the species list and populations of an island, since a rising intertidal "ring around an island will shrink the intertidal habitat. The comparable effect on the nearshore community would be slight, since the intertidal "line" simply shifts up or down but remains nearly the same length. Because of this high sensitivity, the community at Cordell Bank could be expected to fluctuate much more rapidly in time than would the reservoir. The approximation made here is that there is a large fixed pool of organisms from which the local community is selected. The organisms themselves are assumed to be genetically stable; it is their numbers, but not their nature, that are determined by the local environment.

It is valuable to keep in mind several characteristic times and rates: the diurnal cycle, the lunar cycle, the time for an organism to reach maturity and reproduce, individual lifetimes, propagation times, seasons, climate changes driven by astronomical precessions, rates of continental drift and tectonic rates and intervals, rates of genetic mutation, and so on. Whenever two or more such times or rates are comparable, there is the likelihood that the corresponding forces will compete, or interfere, in the community, producing new and interesting relationships.

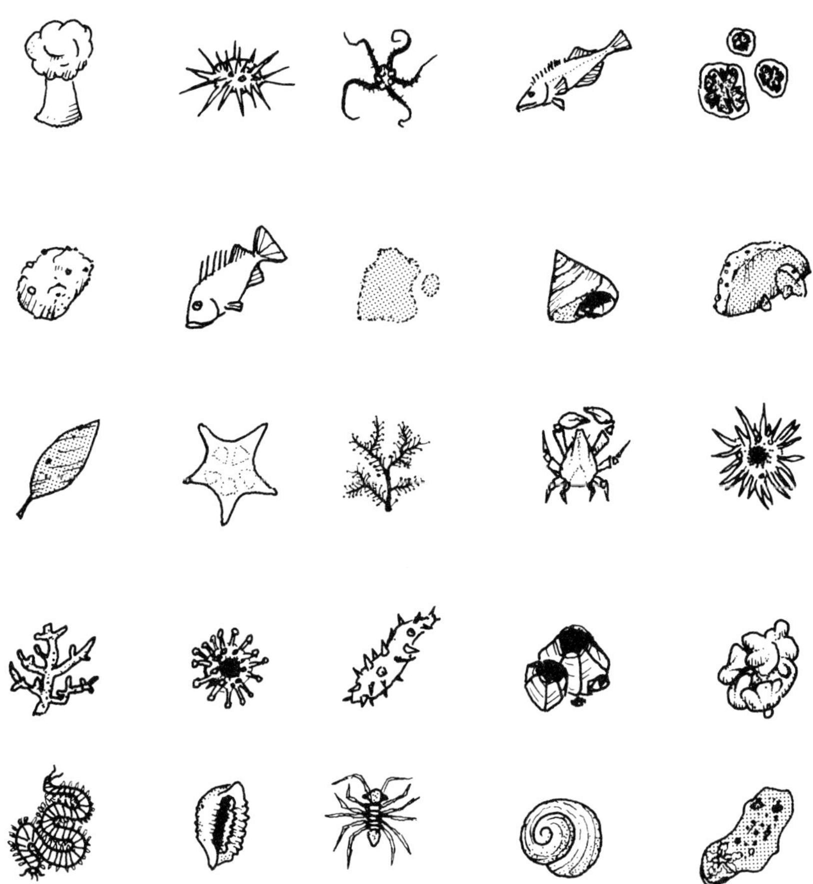

A Place to Be

The biggest limitation at Cordell Bank is, of course, living space. With plenty of food and energy, the item in shortest supply is, quite literally, a "piece of the rock." Desperate for someplace to call home, the organisms pile on top of one another in unexpected ways. For instance, the strawberry anemone *Corynactis californica* is sometimes content to live on sponges, rather than on the hard rocky substrate it prefers, and it forms beautiful spherical colonies rather than the flat sheets familiar elsewhere. The sponges themselves pile on top of one another, at times intertwining with other species of sponges and with tunicates. Hydroids, particularly *Garveia annulata*, which looks like a yellow weed, routinely establish themselves on the very top of these stacks of sponges. The environment on the Bank gives new meaning to "Space Wars"!

SOME ORGANISMS COMMON AT CORDELL BANK

First row:		
	White-plumed anemone	*Metridium senile*
	Giant sea urchin	*Strongylocentrotus franciscanus*
	Brittle star	*Ophionereis annulata*
	Lingcod	*Ophiodon elongatus*
	Orange-red solitary coral	*Balanophyllia elegans*
Second row:		
	Massive yellow sponge	*Halichondria panicea*
	Yellowtail rockfish	*Sebastes flavidus*
	Encrusting red alga	*Cruoria profunda*
	Purple ringed top snail	*Calliostoma annulatum*
	Gray moon sponge	*Spheciospongia confoederata*
Third row:		
	Large-bladed brown alga	*Desmarestia tabacoides*
	Leather star	*Dermasterias imbricata*
	Hydroid	*Garveia annulata*
	Masking crab	*Loxorhynchus crispatus*
	Yellow anemone	*Epizoanthus scotinus*
Fourth row:		
	California hydrocoral	*Allopora californica*
	Strawberry anemone	*Corynactis californica*
	California sea cucumber	*Parastichopus californicus*
	Barnacle	*Balanus nubilus*
	Sponge	*Geodia mesotriaena*
Fifth row:		
	Polychaete worm	*Nereis eakini*
	Pink snail	*Pedicularia californica*
	Isopod	*Munna spinifrons*
	Dwarf turban snail	*Homalopoma luridum*
	Sea lemon	*Anisodoris nobilis*

Hydrocoral and Friends

One of the more successful organisms on the Bank is the pink California hydrocoral *Allopora californica*. It is closely related to the more familiar encrusting purple hydrocoral *Stylantheca porphyra* that forms colorful sheets in relatively shallow water [Morris, Abbott, and Haderlie, 1980]. At Cordell Bank, however, and on other bank tops as well, the spectacular tree-like branching colonies of *Allopora* dominate. This structural form developed to enable the organism to reach above others competing for the same space; the same happened with terrestrial trees. With a relatively small anchor and stem, such branching organisms are able to gain access to a much larger "airspace."

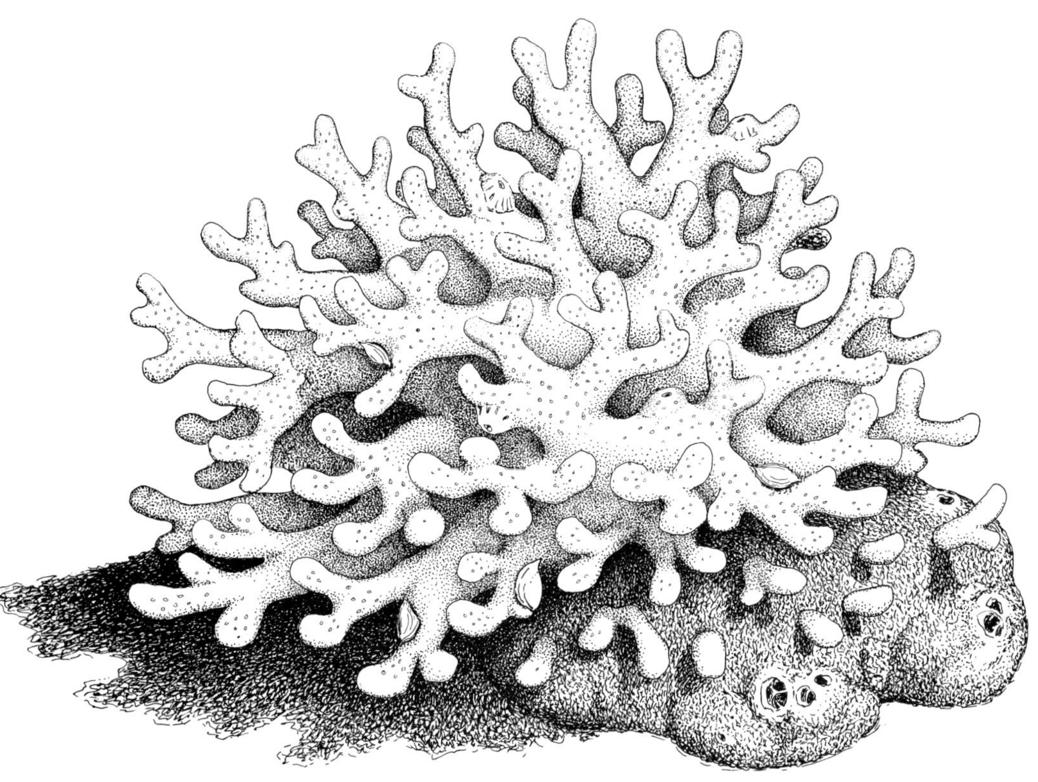

The California hydrocoral *Allopora californica*. This colonial organism has roughly the strength of plaster; it is brittle and easily crushed. It grows at roughly 1 cm per year. Like a tree in the forest, this organism provides shelter, support, and food for many other plants and animals, some of which have evolved into obligate commensals—they must live on the hydrocoral. Here we see the pink snail *Pedicularia*, the barnacle *Armatobalanus*, and the polychaete *Polydora* living within the branches.

Allopora almost certainly did not evolve to its present form at Cordell Bank. Most likely it had already evolved long before the present conditions on the Bank were established. Rather, *Allopora* finds the environment there just to its liking—a hard substrate on which to anchor, and a continuous current carrying a heavy load of tasty morsels. Because it is so well-suited for this environment, it flourishes: the Bank is awash with *Allopora*.

The pink snail *Pedicularia californica* on its hydrocoral host. After a nomadic period, the snail has settled down to a specific location and grown its shell to exactly conform to the shape of the hydrocoral. If the snail were removed, a shallow scar in the hydrocoral would be seen.

Allopora supports several other animals that have evolved into obligate commensals: they live on *Allopora* and can live nowhere else. These include the attractive pink snail *Pedicularia californica*, the small barnacle *Armatobalanus nefrens*, and the polychaete (worm) *Polydora alloporis*. The snail looks like a small pink-purple seed, and roosts in the hydrocoral tree like a bird. The barnacle allows itself to be almost totally encased by the growing hydrocoral, ending up inside a nodular cell with a tiny opening through which it extends its claw-like cirri to rake in food. The worm also lives inside the hydrocoral; its whiplike antennae reach out through a pair of tiny dark holes that look like snakebites.

In addition to acting as host to these commensals, *Allopora* represents a far greater resource to the rest of the community: living space. Its branches generate a relatively large amount of new surface area that is quickly occupied by various colonizers. Normally *Allopora* fends off these intruders, but occasionally the aggressive sponges and tunicates get the upper hand and completely overwhelm the colony, bringing with them all the organisms that can live on sponges or tunicates but not on *Allopora*. The result is a large knoblike projection sticking up from the ridge, as much as 1 m high—gaudy gardens on pedestals, bursting with hangers-on. These knobby projections are characteristic of the landscape at Cordell Bank, prime evidence that the currency on the Bank is a simple foothold.

Algae also attach themselves to *Allopora*. A new species of the red alga *Fosliella*, an undetermined genus of the red algal family Corallinaceae, and an undetermined genus of the brown algal order Ectocarpales were found on the hydrocoral [Silva, 1981]. For *Fosliella*, it is the first time the genus has been observed living on an animal rather than on another plant or directly on the substrate. Cordell Bank is very crowded, indeed.

Sponging

The real space hogs on the Bank are the sponges. With relatively fragile, low-density tissues, they require both a large area for anchorage and a large volume for filtering the water. They thrive best in flowing water carrying a heavy load of microscopic organic matter: protozoa, bacteria, dinoflagellates, and other particles, many too small to be seen with a light microscope [Barnes, 1974]. But this is just what Cordell Bank offers, and the sponges respond with vigor. With ample food supply and few predators, they burst forth in such numbers that they totally dominate the scenery. It's a sponge heaven—probably 90% of the area is covered with sponges.

A typical sight at Cordell Bank: an aggregate of animals piled on each other to form a knobby structure. The structure may be built on a core of the hydrocoral *Allopora*, which provides mechanical support, like a tree. Eventually it is entirely enclosed by the hangers-on. Here various massive subglobular sponges clonal colonies of the strawberry anemone *Corynactis californica*, and tunicates are topped by a colony of the hydroid *Garveia annulata*.

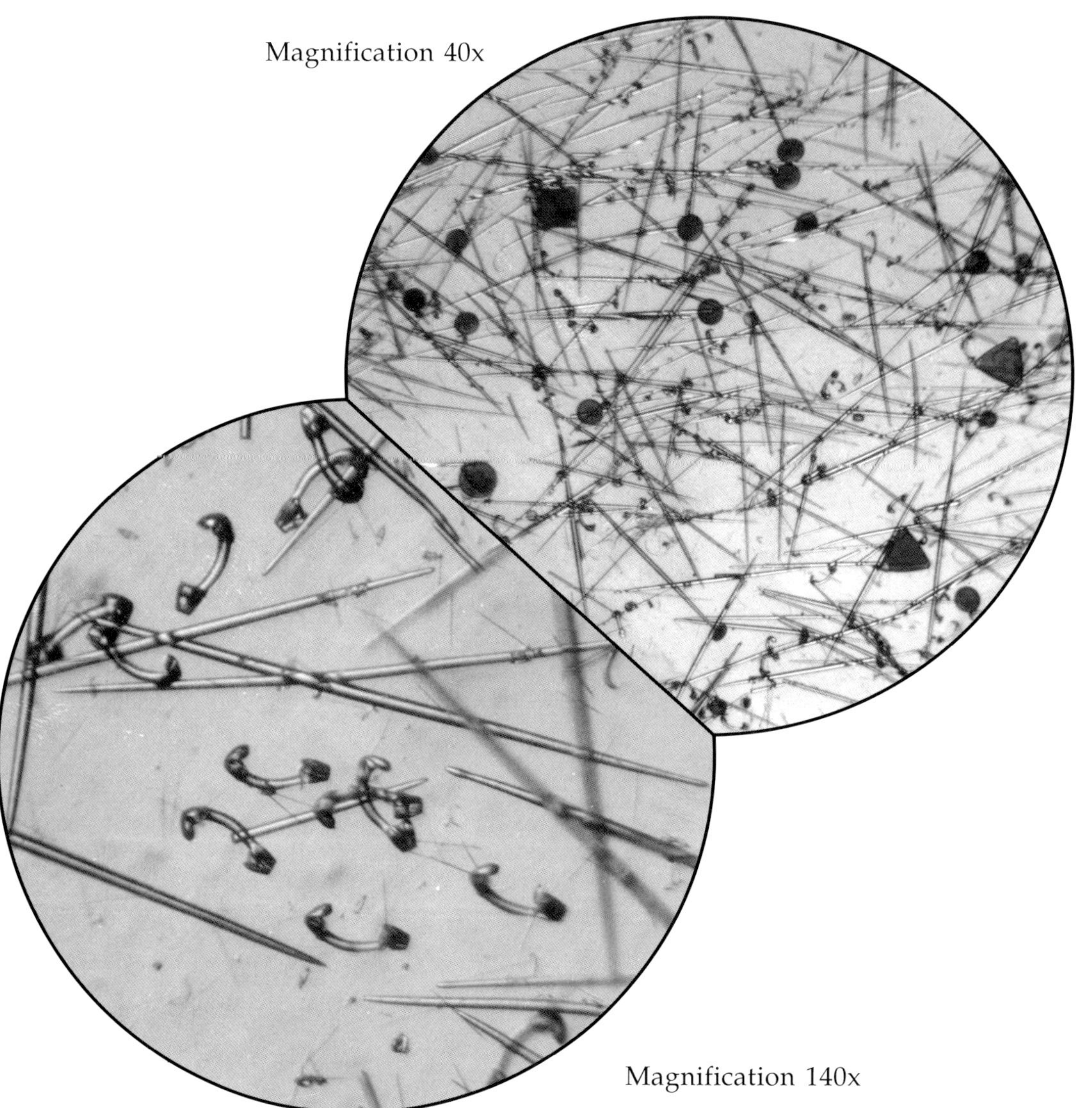

The skeletons of sponges. The tissue of sponges is held together by a thicket of microscopic siliceous needles, hooks, barbs, and pins called spicules. These photomicrographs show some of the bizarre forms spicules can take. The names of various spicule types are almost as bizarre as their shapes: styles, strongyles, tylostyles, acanthostyles, oxeas, anisochelas, cladotylotes, anatriaenes, spherasters, triradiates, sigmas, toxons, forceps, and calthrops, to name but a few. Each species of sponge generates its

Room with a View

There is a very big difference in value between horizontal and vertical living exposure. "Room with a view" is eagerly sought, but only if the view is up. Vertical walls and cliffs are generally very sparsely populated (although some vertical exposures at Cordell Bank are covered with large blankets of *Corynactis*), while adjacent horizontal surfaces are buried under a cover 20 cm or more thick.

Several factors operate to produce this rather spectacular difference. The first is that current-carried propagules (larvae, gravid females, etc.) will settle and successfully colonize a horizontal surface, but won't settle on a vertical surface. This factor is not unique to Cordell Bank, but it probably accounts for the relative absence of sessile, nonphotophilic animals such as sponges, anemones, and tunicates on vertical walls on the Bank.

Another factor is that although many mobile animals such as crabs and gastropods can successfully climb on a vertical exposure, there is increased danger of losing their grip and falling. This factor is also present in other areas, but in the high-current environment on Cordell Bank it probably has a much greater relative importance. Any small invertebrate foolish enough to let go for a moment will most likely be swept away, almost certainly to its death from exposure to high temperatures in water near the surface, desiccation on a nearby shore, starvation on dark desert-like mud at abyssal depths, or digestion in the stomach of a fish.

A third factor involved in horizontal and vertical exposure, and possibly the most important at Cordell Bank, is that most of the light reaching the bottom comes directly from the surface. Horizontal surfaces present a full face to this radiance, but vertical surfaces receive only the relatively small fraction that is scattered and diffused. That the direct illumination is significantly greater than the diffused light is clearly seen in available-light photographs. Light is important because it allows algae to grow, which in turn feeds herbivores, such as some mollusks, echinoderms, and arthropods, which in turn provide food for the carnivores.

(continued from p. 52)
own characteristic set of spicules, and usually it is necessary to study the spicules with a microscope to identify a sponge with certainty. For instance, the sponge *Lissodendoryx firma* has styles, strongyles, sigmas, isochelas, and forceps. Identification of these organisms clearly is a job for a specialist! Several of these photographs show round, square, and triangular diatoms entangled in spicules of the sponge.

From Bottom to Top

The variation of the plants and animals with depth is one of the most intriguing aspects of Cordell Bank.

At depths greater than 70 m, most of the bottom is covered with coarse gravel sediment. This sediment is almost 100% calcareous (calcium carbonate). It is composed of the sifted, ground-up remains of gastropod and bivalve shells, sponge spicules, urchin and foraminiferal tests, coral skeletons, and miscellaneous pieces of arthropods, fish scales, otoliths, opercula, and worm tubes. Within the sediment lives a variety of animals, mostly tiny arthropods, worms, and protozoa. Wandering around the surface of the sediment are brittle stars, crabs, and urchins. Often large aggregates of rockfish hover just above the sediment.

Rocks that protrude up out of this sediment, and the lowest portions of the steep cliffs and ridges, are very thinly covered. Among the most common organisms on these rocks are solitary *Corynactis californica* (in shallower water it lives in clonal aggregates), stony corals (*Balanophyllia*, *Caryophyllia*), small colonies of hydrocoral (*Allopora*), and a few small globular and encrusting sponges (*Polymastia pachymastia*, *Acarnus erithacus*). A few erect red algae (*Maripelta rotata*, *Polyneura latissima*, *Callophyllis* spp., *Fauchea* spp.) stand out conspicuously because of the paucity of invertebrates. Every "bare" area (about half the available area for depths around 60 m) is covered with red or pink encrusting algae (*Cruoria profunda* and others).

Between about 60 and 50 m, the size of the *Allopora* colonies increases to about 10-20 cm. Several large animals appear as isolated individuals, including the larger anemones such as *Metridium senile*, the sea cucumber *Parastichopus californicus*, the sea stars *Mediaster aequalis* and *Dermasterias imbricata*, the urchin *Strongylocentrotus franciscanus*, and several white, almost translucent tunicates. Lingcod are common here. The cover is dominated by invertebrates, and is nearly total, although it may be thin in places.

Between 50 and 45 m, the cover increases to virtually 100%, and its thickness is typically 10-15 cm. No bare rock shows. The sponges are tightly packed, forming a continuous carpet, but not usually stacking on top of each other or on other organisms. Decorator crabs and several species of the top shell *Calliostoma* crawl about everywhere. Algae are scarce. The wheel-like gray moon sponge *Spheciospongia confoederata* forms large cross-buttressed structures. There are few urchins, but there are huge amounts of *Corynactis* and *Allopora*. In this region, and everywhere above, the beautiful rosy rockfish *Sebastes rosaceus* is abundant, showing its brilliant orange skin and six white spots.

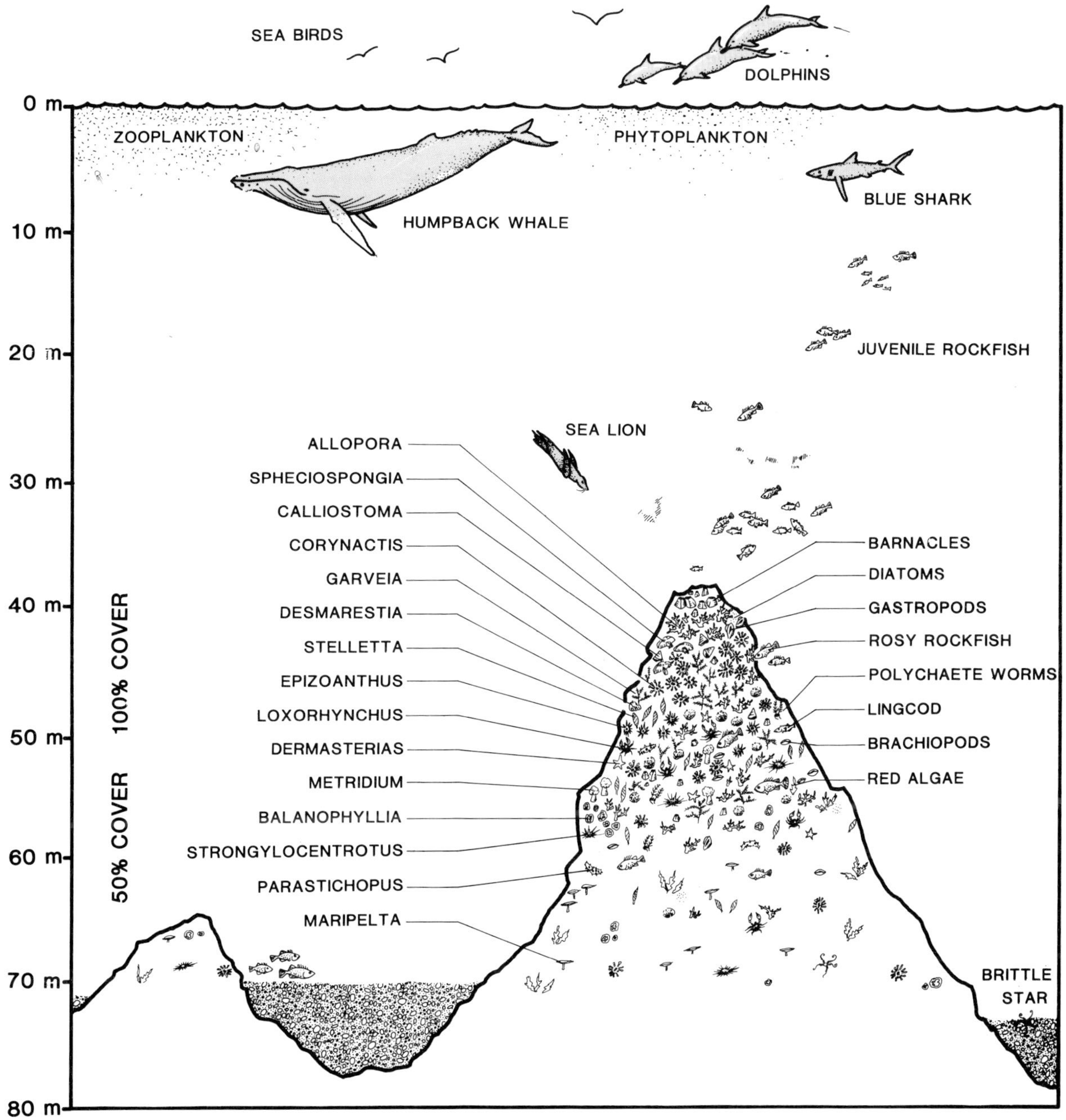

Schematic of the variation with depth of the community on Cordell Bank. The shallowest areas, above about 40 m are covered 100%; there are no bare areas. A few common organisms are shown in roughly the observed sequence. Some animal groups, such as brittle stars, have great depth range; they are found at all depths. Others, such as barnacles, prefer to live only on the shallowest points, and are found within a narrow cap at the tops of pinnacles. As death claims an animal, its undigestible parts such as a shell, test, spines, skeleton, or operculum fall into the sediment catchments, accumulating over thousands of years to form deposits many m thick. To the extent that these deposits are not vertically mixed, they preserve a stratified record of the populations over time.

Above 45 m, competition for space is desperate. Very large colonies of *Allopora*, up to 30 cm across, reach out from the ridge tops. Various sponges and the lobed tunicate *Cystodytes lobatus* use the hydrocoral for mechanical support. Other sponges, including *Stelletta clarella*, crowd each other. The anemones *Corynactis* and *Epizoanthus* form spherical colonies. The hydroid *Garveia* attaches itself to the very tops of knoblike piles of sponges. Except on the tip of the shallowest ridge, the cover above about 45 m is practically homogeneous—there is no obvious variation of the community with depth.

On the very shallowest point, on a ridge at the north end, is found a cap covered almost solidly with barnacles. Attached to the barnacles is a profusion of bright red blades of algae, covering perhaps a third of the area. Patrolling this prickly garden is a corps of decorator crabs, *Loxorhynchus crispatus* and *Scyra acutifrons*, and the inevitable top shells, especially *Calliostoma*. About 3 m below the tip of the ridge the cover suddenly (within the vertical space of 1 m or so) changes to the "normal" complex of anemones, hydrocorals, hydroids, and sponges.

We have gone into some detail about the depth variation of the community at Cordell Bank, principally because it is unique. Although almost all the animals and a few of the algae found on the Bank are commonly found elsewhere, the particular combination and variation with depth is found nowhere else. It is the combination of rocky ridges and pinnacles, strong current, and exceptionally clear water that establishes this unique community, and that fortuitously makes it beautiful and scientifically interesting.

(Left) The barnacle *Balanus nubilus*. A tightly packed colony of barnacles was found on the tip of the shallowest pinnacle on Cordell Bank, about 36 m deep. This colony succeeded in displacing almost all other species, giving the tip a jagged texture that was brilliant white. The colony ended abruptly about 3 m deeper, where anemones and sponges were dominant. (Right) The barnacle *Mega- balanus californicus*, which also forms rather spectacular clusters.

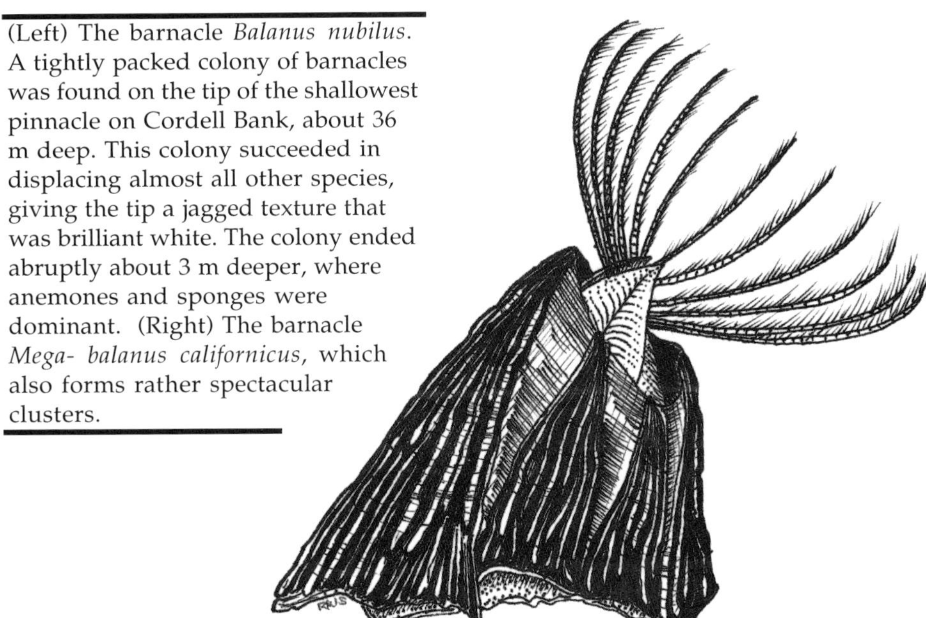

It is noteworthy that the general causes of the vertical zonation at Cordell Bank are opposite those of intertidal communities, if we ignore topography as a variable [Carefoot, 1977]. For intertidal communities, the upper limits are set by physical factors (mechanical stress, desiccation, temperature, exposure to ultraviolet radiation), while the lower limits are determined by biological factors (competition, predation). In contrast, for subtidal communities such as that at Cordell Bank, the upper limits are determined by biological factors (competition, predation), while the lower limits are determined primarily by one physical factor: the light level. Below the depth above which the community is sensibly homogeneous, about 45 m, the decreasing intensity of the light with depth is the primary cause of the reduction of populations.

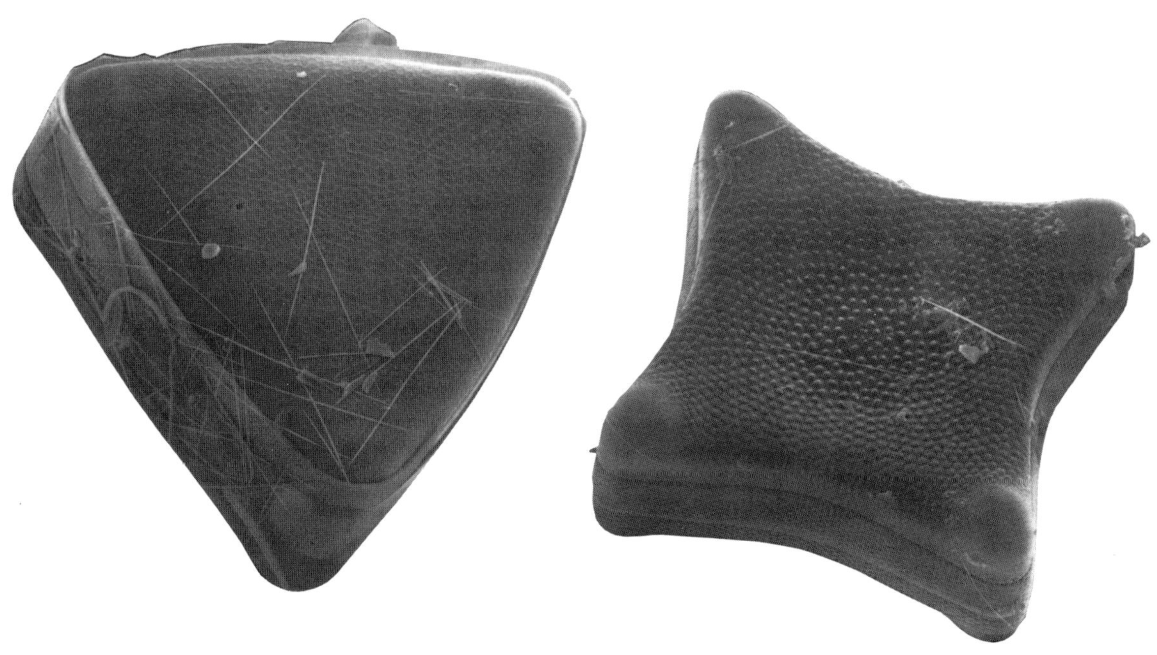

Two forms of the same species, the common diatom *Triceratium*. The origin of the name is easy to understand from the left photograph, but the four-cornered specimen on the right is the same species. It apparently results from a slight genetic imperfection. These photomicrographs were made with a scanning electron microscope. The needle-like debris are sponge spicules. (Magnification 450x).

Study of a rare diatom, *Entopyla* cf. *E. incurvata*. This diatom, common at Cordell Bank, is found in only a few widely separated places in the ocean, which is taken as evidence that it is relictual. Formerly it was more widely distributed, but as conditions changed, it was driven to extinction almost everywhere, leaving a few isolated pockets where the sequence of conditions was favorable for its survival. The last description of this microscopic plant in the scientific literature was in 1896; these scanning electron photomicrographs show the details of the structures for the first time. The diatom shown here is about the size of the smallest spot that can be seen by the unaided human eye, just a speck. These photos show, however, how complex and subtle the genetic engineering has been, in even these minute organisms. (Magnification 160-4500x).

Energy

The fauna decrease with depth even faster than do the flora. This occurs because, as we go deeper, the dwindling available energy becomes an increasingly precious commodity, and those organisms that use it most efficiently compete for it more successfully than less efficient organisms. Since animals don't utilize the light energy directly (they do not photosynthesize, and heat input is negligible in the deep environment at Cordell Bank), they obtain their energy exclusively from things they eat. Animals, especially mobile animals, must expend a significant fraction of their energy chasing, capturing, or prospecting for food [Calow, 1981]. Although predators do consume other animals, the ultimate source of energy for all the animals is the plants. Thus, a fraction of the total plant mass is consumed by animals and used in the effort of moving around. But this fraction is wasted, since the energy it represents is expended as heat generated in muscle activity and carried away in the cold water, rather than in making more animals. Plants have little such waste; all the energy they obtain in excess of the very small amount used in respiration goes to making more plants. In the dimness of the depths, efficiency counts. The plants win.

If we assume that the most successful organisms are those that utilize the light most efficiently (directly or indirectly), we can predict the decrease of various classes of benthic organisms with depth: first the green algae lose out, then filter feeders that depend on phytoplankton, then brown algae, then the large predators, then the large herbivores. Finally, the only animals that can eke out a living are small herbivores, meiofauna (animals living in the sediment), and a few highly specialized forms. The algae, increasingly free of their animal consumers, increasingly grow to the limits set by the available energy and substrate area.

Within relatively broad limits, this sequence is observed at Cordell Bank. The animal cover thins out sufficiently near 60 m that the algae become a prominent part of the landscape. Below 75 m even the algae are sparse. This is, incidentally, far from the maximum depth at which plants can live; crustose algal forms recently have been collected from 268 m [Littler, et al., 1985].

It might be noted that although the mechanical energy associated with subsurface currents is not used directly by organisms, it is a crucial driver for the distribution of food and the spread of propagules. It therefore is a major factor in competition: organisms that take advantage of this "freebee" tend to occur in sessile colonies that wait for materials to drift by, while mobile hunter/gatherers are generally solitary individuals with less patience and more mechanical energy.

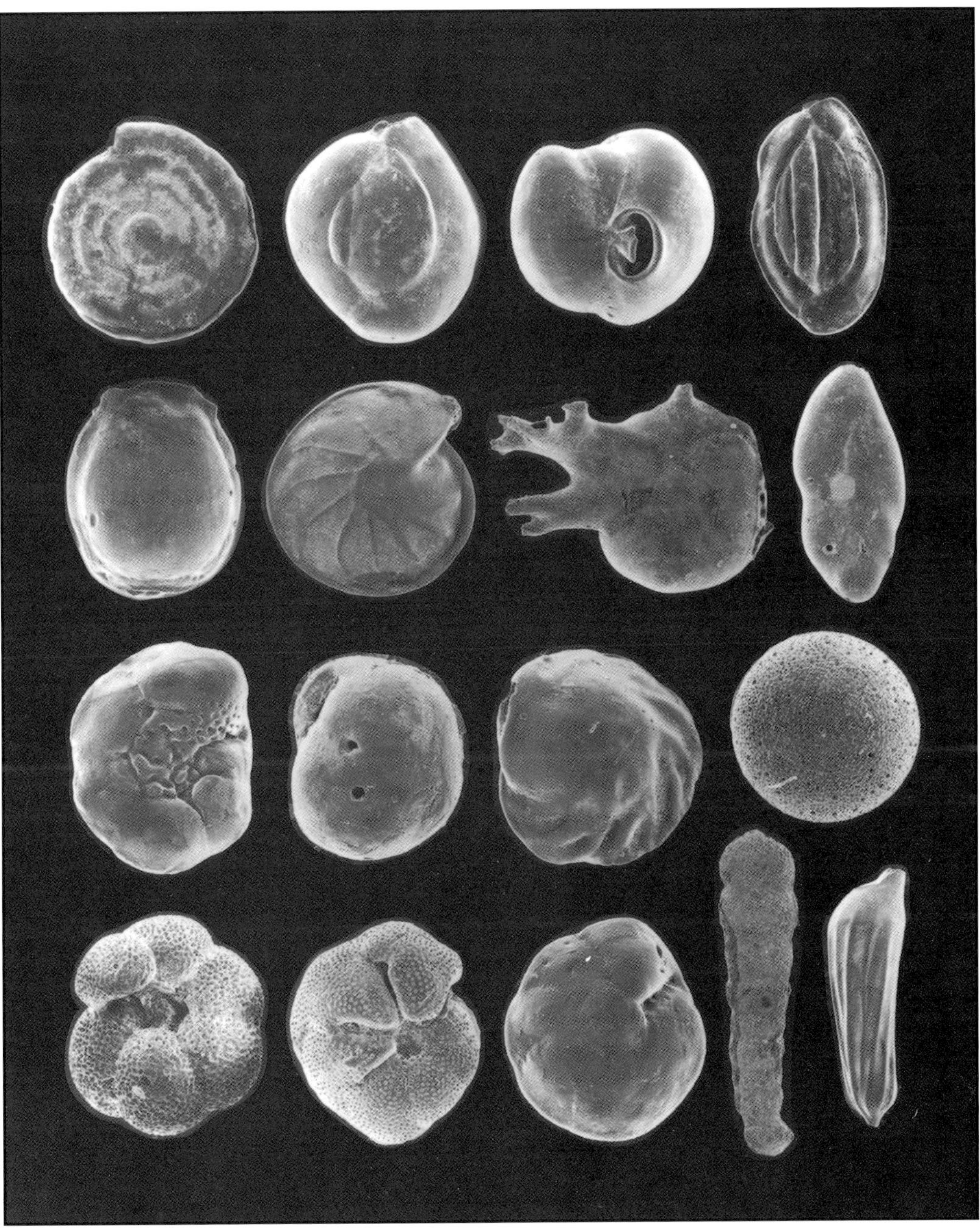

A selection of foraminifera found in the fine sediment at Cordell Bank. These animals are sometimes so numerous they form large deposits of mud. Around Cordell Bank there are areas of green mud that is almost 100% foraminiferal ooze.

(Above) A selection of small gastropods collected at Cordell Bank. This collection includes specimens that were alive when collected, some that were long dead and mixed in with the coarse gravel sediment, and some that show evidence of having been ripped apart by crabs or drilled by other gastropods. (Below) A comparison of two small typical gastropods (left) and two typical foraminiferans (right), each showing prominent spiral structure. In spite of the similarities the animals are not even remotely related; the gastropod is a multicelled animal, while the foram is single-celled.

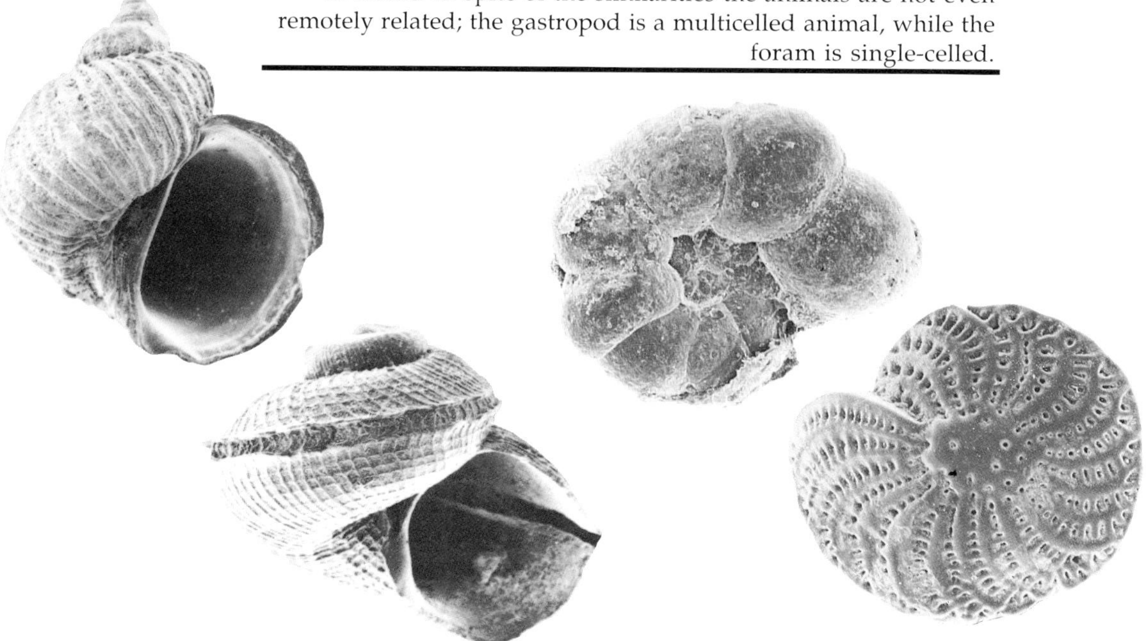

Seen in Different Light

One may ask whether the relatively high light penetration at Cordell Bank simply shifts the populations deeper without changing the *composition* of the community. The answer must be "no," since although the light level controls the local rate of photosynthesis, and therefore the depth at which various attached plants can exist, the energy used by the community is not solely derived from the light through local photosynthesizing plants. Significant energy is derived from nutrients transported to the Bank by currents and upwelling. Thus, as we increase the light penetration, some organisms (particularly plants) will extend down to new depths corresponding to the same threshold light levels, but other organisms will not respond by moving deeper. This differential shift will alter the quantitative relationships between herbs and herbivores, prey and predators, and so on. A new equilibrium will be established that will have a new balance among the various species, i.e., a new community.

The intensity of the light not only affects the large attached algae (seaweeds), but also the phytoplankton. These mostly microscopic, free-floating algae in turn determine the distribution of the zooplankton. To the extent that the water flows without vertical mixing, this plankton layer is dense near the surface and falls off with increasing depth. As the water carrying this stratified soup passes over the Bank, the filter feeders (such as sponges) and small predators (such as anemones and small arthropods) living on the shallowest points dine in splendor, while deeper denizens go hungry.

The color of the light is also crucial to the community. Not only does the light lose its intensity as it penetrates the water column, it also loses its highs and lows—the reds and violets [Tyler and Smith, 1970]. What's left is a paltry amount of mostly blue-green light. Most green plants can't absorb this light efficiently, so they just can't live at great depths. Red algae, on the other hand, can absorb the blue-green light and carry on photosynthesis, albeit slowly. As a result, practically all the plants on Cordell Bank are red algae.

The fact that there is only blue-green light at great depth accounts for the fact that many organisms that are red in daylight appear almost black on the bottom. The red plants efficiently absorb blue and green light; little is reflected.

The Red Light Mystery

The striking observation of the red color mentioned earlier (page 35) was one of the mysteries presented to the first divers on Cordell Bank. The mystery was finally solved in 1983. Using high-speed color film balanced for daylight, divers [Dvorak, 1983; Kruse, 1984] took photographs on the bottom using the available light (that is, without a strobe). When the film was processed and printed normally, great blotches of red color appeared, separated by dark green and black. The red color came from a well-known organism, the strawberry anemone, *Corynactis californica*. The explanation: *Corynactis* is fluorescent!

Fluorescence is a physical process in which an object absorbs light of one color and emits light of another color. For instance, an object being illuminated in blue or green light might emit red or orange light. This blue/red fluorescence is common to many organic substances; not surprisingly it is sometimes found in living organisms. In clear oceanic water, about 40-50% of the blue light (wavelength 435 nm) and about 10% of the green light (570 nm) incident on the surface penetrates to 50 m [Nicol, 1960]. The intensity of red light (650 nm) is less than 0.1% of the surface intensity. In this light, organisms that are red or pink in daylight will appear black. Organisms that are blue in daylight will appear blue underwater, but there aren't many of these at Cordell Bank. Purple organisms will appear purple on the bottom, and there are some of these, particularly some colonies of the California hydrocoral *Allopora californica*. But only organisms that fluoresce red will actually appear red to an observer on the bottom.

The fluorescence of *Corynactis* was confirmed in the laboratory in 1984. Specimens collected from the Bank were brought to an aquarium and illuminated with blue, violet, and ultraviolet light. They lit up like red Christmas tree lights, glowing with almost the same bright red color they exhibit in daylight. Different colonies showed somewhat different patterns of fluorescence. Just what advantage this lovely trait gives the anemone in its natural environment is open to debate. Among the specimens collected from Cordell Bank, only *Corynactis* was found to be strongly fluorescent. *Corynactis*, however, is not alone in the biofluorescent world. Many other anemones and other animals, including the terrestrial scorpions, glow like minerals under ultraviolet light [Zahl, 1963; Powell, 1964]. Well after we had solved this mystery for ourselves, we discovered that it had been observed by other divers more than 25 years earlier [Limbaugh and North, 1956].

The strawberry, or club-tipped, anemone *Corynactis californica*. This delicate animal is one of the most common off the California coast, and is seen in large numbers at Cordell Bank. It reproduces by direct fission, thereby forming colonies of genetically identical individuals. Its habit is to cover large vertical walls with broad, flat carpets of individuals. At Cordell Bank, however, it is commonly seen forming spherical aggregates, as shown here. These aggregates may form around a protruding hard substrate, such as a small hydrocoral colony, perhaps a response to the scarcity of available rocky substrate.

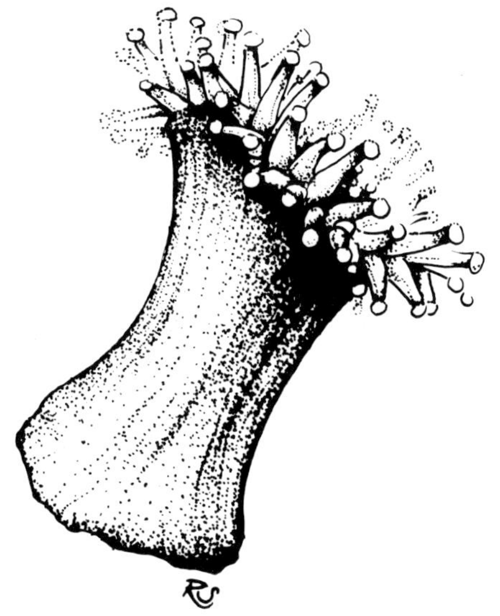

Going with the Flow

Another environmental factor important in determining the distributions of the organisms is the flow of water over the Bank—the currents. These were described in the previous chapter where it was indicated that the flow is very complicated. Upwelling, turbulent eddies and jets, and reverse currents are but part of the story. All this water movement significantly affects the biota [Owen, 1980]. Eddies move slowly, allowing higher concentrations of organisms and nutrients to be maintained for longer times. Jets move fast and transport significant amounts of heat, nutrients, biota, and sometimes pollutants.

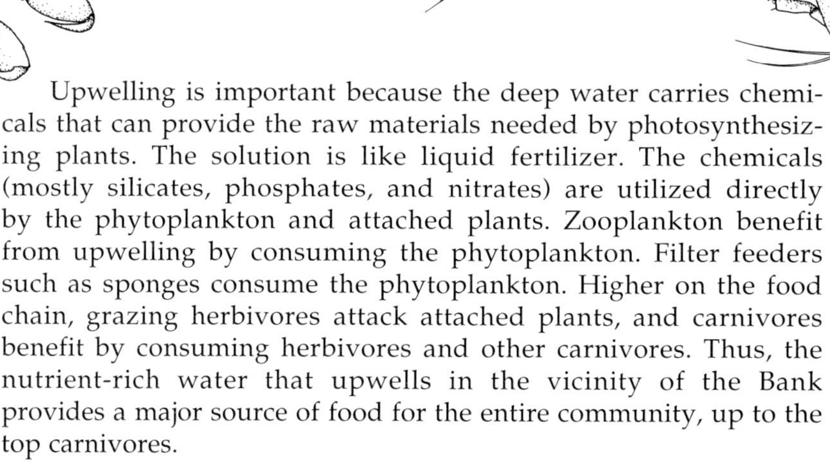

Upwelling is important because the deep water carries chemicals that can provide the raw materials needed by photosynthesizing plants. The solution is like liquid fertilizer. The chemicals (mostly silicates, phosphates, and nitrates) are utilized directly by the phytoplankton and attached plants. Zooplankton benefit from upwelling by consuming the phytoplankton. Filter feeders such as sponges consume the phytoplankton. Higher on the food chain, grazing herbivores attack attached plants, and carnivores benefit by consuming herbivores and other carnivores. Thus, the nutrient-rich water that upwells in the vicinity of the Bank provides a major source of food for the entire community, up to the top carnivores.

Microislands

The horizontal distribution of the biota is also of great interest. We may ask, for instance, whether the biota at the north end of the Bank is essentially the same as, or significantly different from, the biota at the south end. Although any reasonably complete answer to this simple question is going to be complicated, a partial answer can be given. If mollusks are representative of the biota as a whole, their distribution indicates that the north and south ends are "somewhat different." More genera and species of mollusks have been collected than of any other phylum: 145 species in 110 genera. Of these, 22 species were found only in the north, while 8 were found only in the south. These numbers are consistent with the idea that by-and-large the biota is common to all areas on the Bank. Because of statistical fluctuations in populations, however, at a particular time a particular species may not be occupying all of its potential microhabitats. For example, at any given time, a certain species may be present on a particular isolated peak. Later it may disappear from there, but perhaps it will have been established on another peak nearby. It is the barriers between hospitable microhabitats (e.g., the deep water between the peaks) that limits propagation, hence populations.

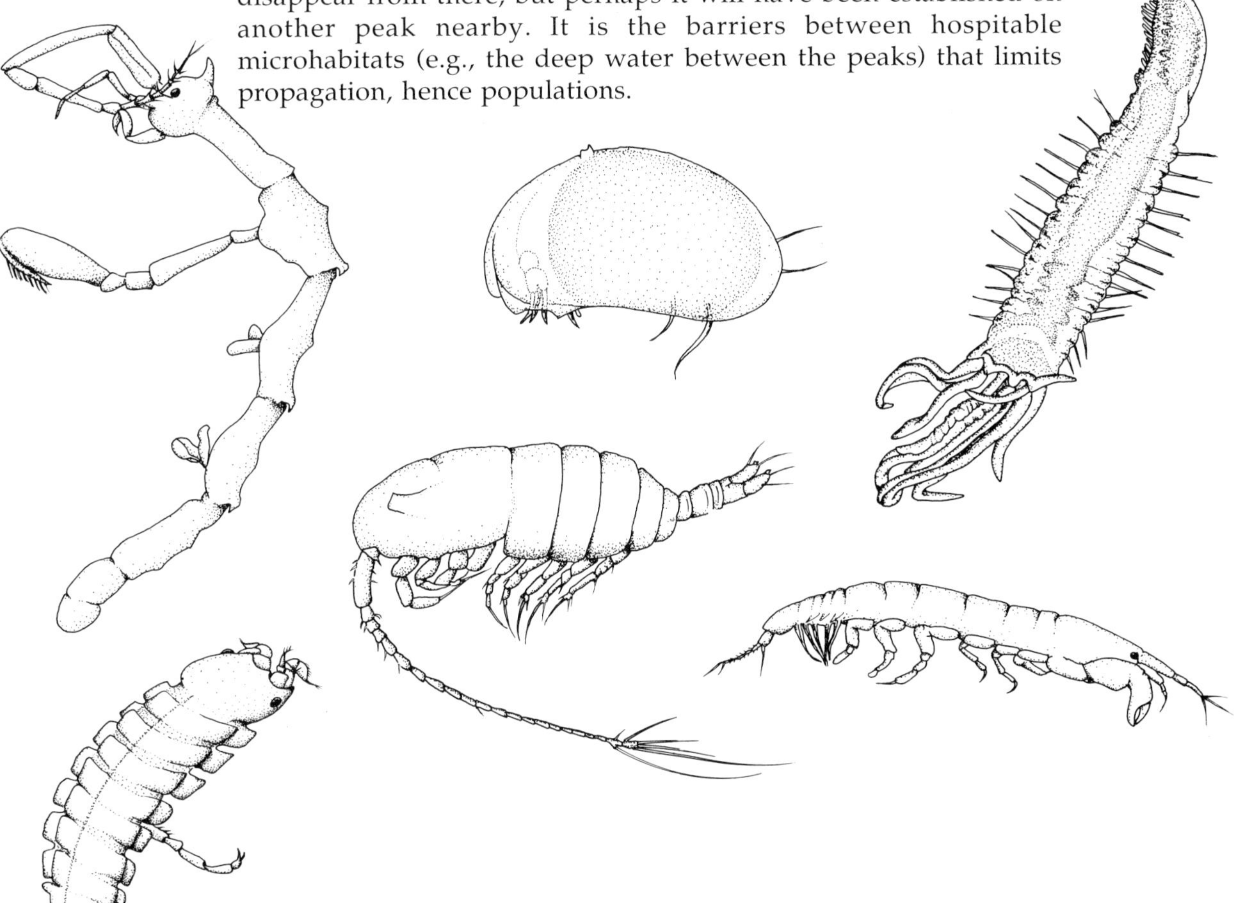

Nothing Remains the Same

So far we have been concerned primarily with static populations: how the present conditions determine the present community. But the variation in time is also interesting. We are immediately confronted with several questions: What seasonal variations occur? Is the community in equilibrium; that is, left alone under present conditions, will it remain the same, or spontaneously change its composition? What is the response of the community to trauma, such as mechanical clearing of a small area or sudden depletion of various species?

Unfortunately, we do not yet have enough information to answer the first two questions. Both questions are, however, amenable to field studies. Determining the seasonal variations simply requires sufficient fortitude to brave the fickle and sometimes fiendish weather year-round.

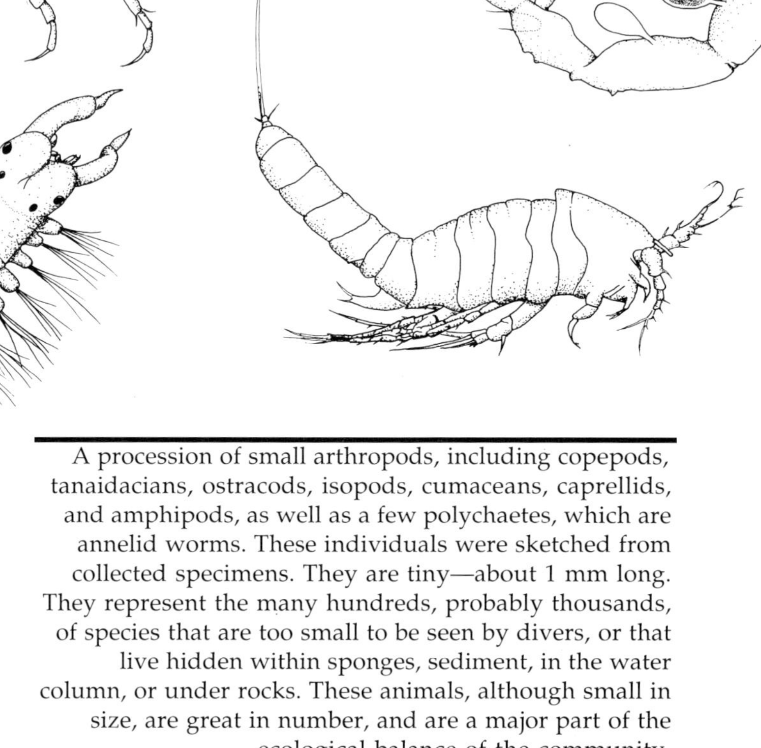

A procession of small arthropods, including copepods, tanaidacians, ostracods, isopods, cumaceans, caprellids, and amphipods, as well as a few polychaetes, which are annelid worms. These individuals were sketched from collected specimens. They are tiny—about 1 mm long. They represent the many hundreds, probably thousands, of species that are too small to be seen by divers, or that live hidden within sponges, sediment, in the water column, or under rocks. These animals, although small in size, are great in number, and are a major part of the ecological balance of the community.

The question of equilibrium is more difficult [Simberloff, 1983]. One way to answer that question would be to just wait and watch, but this not only is boring but also may not give an unequivocal answer since the environment may change along with the community. Another way is based on the fact that equilibrium is established when various rates balance. For instance, in equilibrium, the rate of influx of new species must balance the rate of local extinction of species [MacArthur and Wilson, 1967]. Determining these rates in the field is tantamount to determining whether the community is in equilibrium. Unfortunately, we are as yet a long way from making such determinations on Cordell Bank.

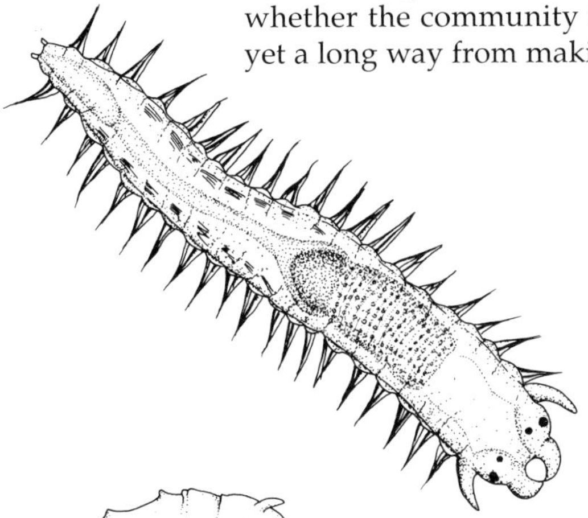

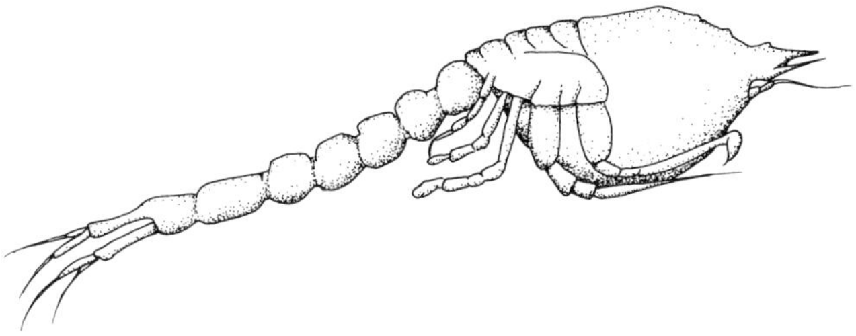

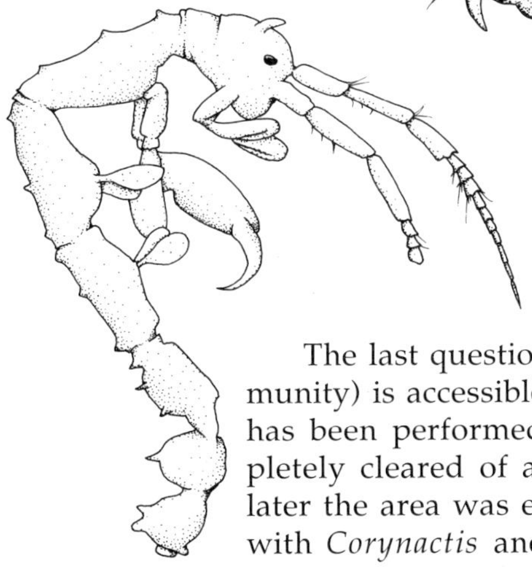

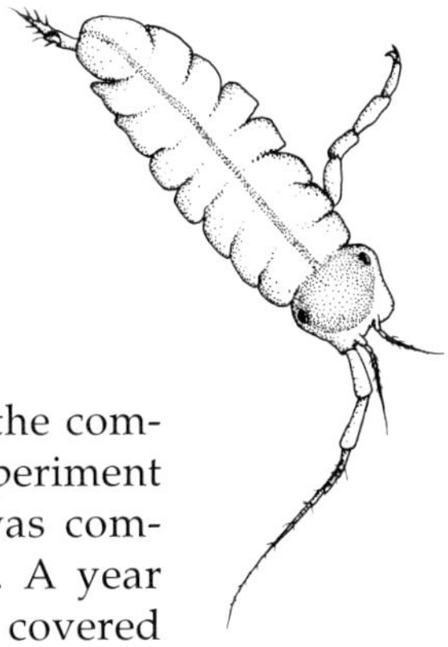

The last question (regarding the effects of trauma on the community) is accessible to experimentation, and one such experiment has been performed. A small area on a shallow ridge was completely cleared of all organisms by scraping them away. A year later the area was examined, and found to be completely covered with *Corynactis* anemones. It seems reasonable to conclude that *Corynactis* is the first colonizer of bare rock, later to be supplanted by other, more persistent, organisms. In a number of photographs of other areas of the bottom, however, large patches are seen covered almost exclusively with *Corynactis* in a manner very similar to the experimental area. Although one interpretation of these photographs is that these areas were cleared by an unspecified agent (presumably biological), and subsequently colonized by *Corynactis*, another possibility is that at those particular locations, *Corynactis* simply competed successfully for the space.

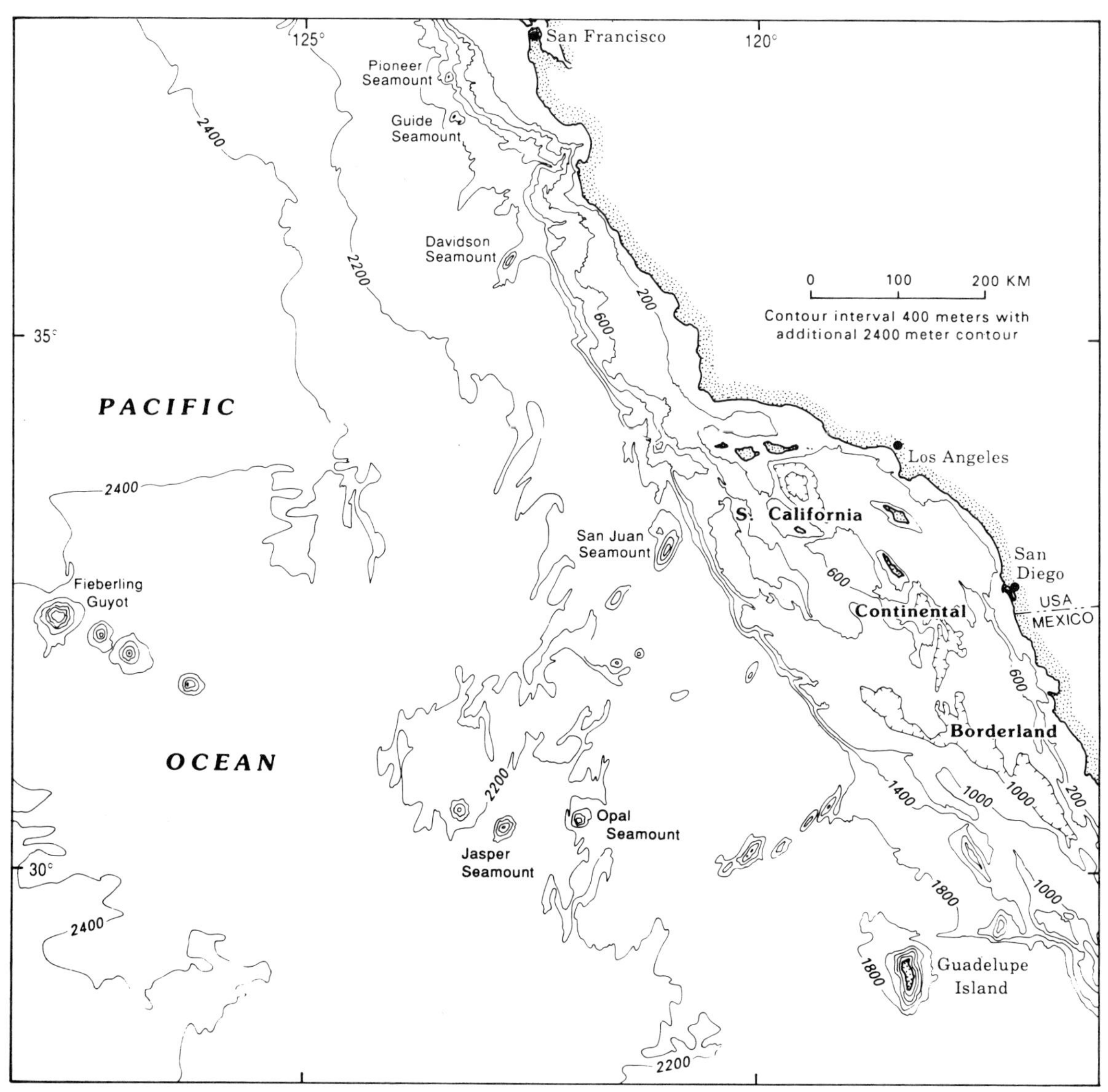

Bathymetric chart of the southern California Pacific. In addition to the Channel Islands west of Los Angeles and Guadalupe Island off Baja, there are many relatively isolated shallow places that could be considered "underwater islands." Communities of organisms usually take advantage of such topographic highs, although the communities on deeper seamounts will be less spectacular than on shallower rocky banks and reefs.

RELATION TO OTHER AREAS

On the Horizon

Cordell Bank certainly *seems* isolated. Lying so far from shore and so deep, it generates the feeling of being remote and unaffected by the nearest land.

However, no island is truly isolated. Cordell Bank lies immersed in a larger ecosystem with which it interacts. Although details are largely unknown, some preliminary information is available that reflect these interactions.

The nearest land to Cordell Bank is the Pt. Reyes head and peninsula, 37 km east. This is the home of hundreds of sea lions, seals, and other large animals. Many of these animals make the swim to the Bank in search of food, and presumably they are generally rewarded. But how do they know there is a lush supply of food beyond the watery horizon? Although they are probably driven at least partially by raw instinct, there is another possible mechanism. The water passing over the Bank is often transported directly to Pt. Reyes, carrying with it the outfall of plants, organic debris, and living animals that signal the presence of a biologically active area upstream. The transport patterns have been observed by experiments in which drifters were released on Cordell Bank, and subsequently found at Pt. Reyes [Conomos, et al., 1970]. Thus, Cordell Bank is like a volcano, spewing forth a plume of particles and chemicals that is readily detected by highly mobile predators such as marine mammals and birds.

Sensing a streaming chemical imbalance in the water is one of the ways that sharks detect potential prey. By following the stream to its source, they find the prey. Interestingly, although Great White sharks are common in the entire Gulf of the Farallones, they have never been sighted at Cordell Bank. It is possible that prey in the open water is just too infrequent for them to bother with. Perhaps they prefer to remain nearer to shore, where the food is easier to find. Recent observations of shark attacks at the Farallon Islands suggest that there is a relatively small population of sharks that migrate, most animals arriving at the Islands in late fall, feeding on young pinnipeds, and then moving on. The author's experience, including many dives at the Farallons, supports this perception.

Island Populations

The Farallon Islands lie some 50 km southeast of Cordell Bank. They comprise Southeast Farallon (the largest island), Middle and North Farallons (craggy rocks), Noonday Rock (a wicked washrock named after the ship it wrecked), and Fanny Shoal (also named after a wrecked ship). All these prominences, together with Cordell Bank, lie in a roughly straight line. Furthermore, they are all composed of roughly the same material, the quartz diorite making up the Salinian Block. Since all these areas are washed in the same California Current system, it is natural to expect many similarities in the biota.

Unfortunately, until very recently there has been very little collection of subtidal invertebrates from the Farallons [FRG, 1978], primarily due to the danger of sharks to divers. During 1990 the author and colleagues carried out a series of expeditions to the North Farallons, and obtained a large collection of algae and invertebrates, but these have not yet been analyzed in detail. It is therefore difficult to quantitatively compare the subtidal populations there with corresponding populations at Cordell Bank at present.

We can speculate, however, that the Islands might in some ways be intermediate in character between the strictly subtidal community at Cordell Bank and the intertidal community on the mainland shore (say at Bolinas Bay or at Bodega Bay). One bit of data that might be brought to bear here is the list of shelled gastropods. Lists of the species that have been found at the four localities indicate that roughly the same number of genera and of species are known from Cordell Bank, the Farallons, Bolinas, and Bodega. Furthermore, the ratio of number of species to number of genera, which is one measure of the completeness of sampling of the communities, is about 1.5 for all four sites. Therefore it may be meaningful to compare these sites.

Detailed comparison of the lists of species identified at each of these four sites shows the following: About half of the species from the Farallons are also found at Cordell Bank, whereas only about a tenth of the species found at Bolinas and about a fifth of those found at Bodega are found at Cordell Bank. In other words, there is significantly greater overlap between the Farallons and Cordell Bank, and between the Farallons and the mainland, than between Cordell Bank and the mainland. The Farallons, as "real" islands, seem to stand in an intermediate position between the mainland and the underwater island, Cordell Bank.

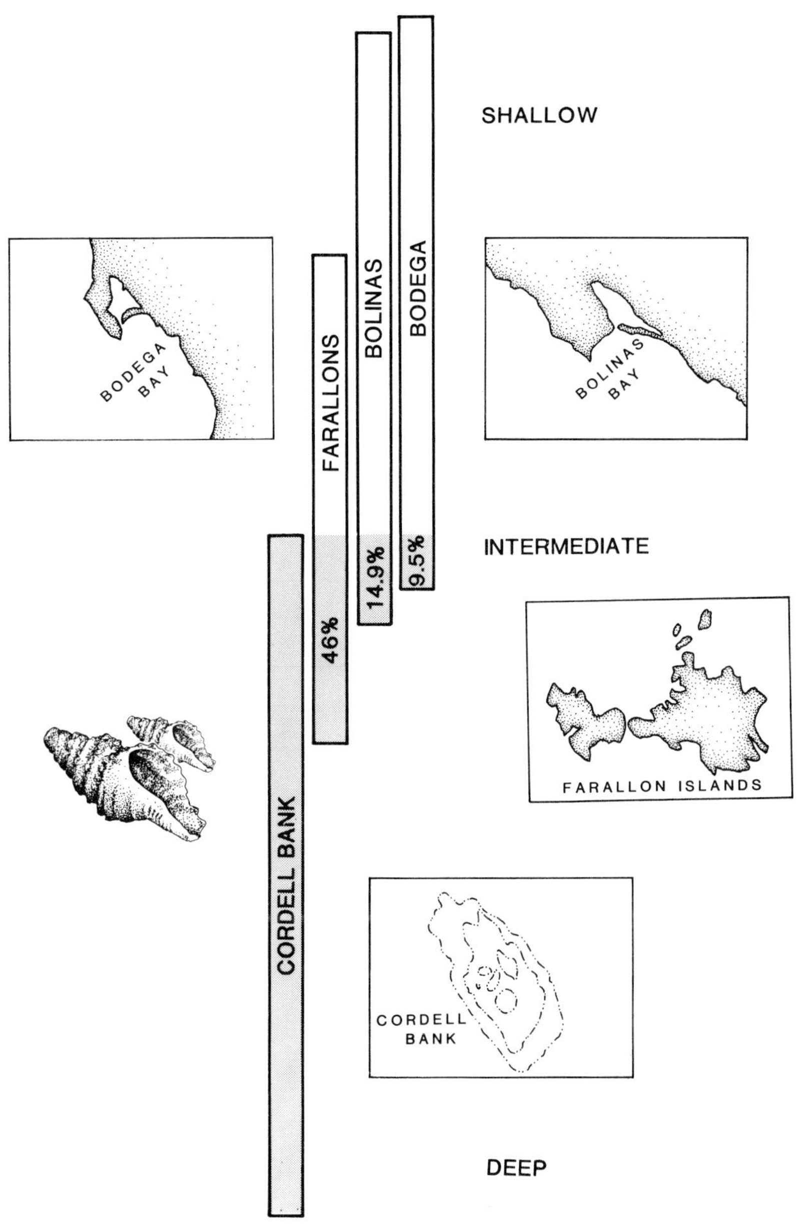

A comparison of the overlap between the shelled gastropods found in four different marine sites comprising three distinct types: Cordell Bank (a deep underwater island), the Farallons (offshore islands), and Bodega Bay and Bolinas Bay (shallow coastal bays). The number of species found at these sites are plotted as vertical lines; their lengths are proportional to the number of species found at each site. The number of species in common between Cordell Bank and each of the other three sites is indicated by the overlap of the lines: the greater number of species in common, the greater the overlap. The arrangement of overlap suggests that the Farallons are, in some sense, intermediate between Cordell Bank and the shallow coastal bays, roughly consistent with our image of an offshore island being intermediate between a shallow and a deep environment.

Taxonomic Selection

In a similar way we can examine the relationship of the populations at Cordell Bank to those of California as a whole. The latter, in some sense, constitutes a relatively large and diverse reservoir within which Cordell Bank is immersed, and with which the Bank exchanges organisms. In this sense, California could be considered a "parent" population to that of the Bank. This comparison might show the operation of selection in establishing the community at Cordell Bank. That is, some factor operating at Cordell Bank may favor or disfavor one or more taxonomic groups of organisms. In the absence of selection, all taxonomic levels would be represented in the same proportions on Cordell Bank as in the larger California population (within the statistical accuracy). If the observed proportions are different, one would suspect that selection is operating.

For this comparison, we examine the algae known from Cordell Bank and from all of California. It must be remembered that deepwater algae generally are more similar worldwide than are the intertidal algae, and that the deepwater algae are not derived from the nearshore community. Thus, the California population is not strictly the parent of the Cordell Bank population. The comparison is nevertheless instructive, as we shall now demonstrate.

The comparison was formulated by A. Kerstein [1985], who phrased it as follows: We know the distribution of species among various taxa of the California population, and we know the fraction of the California species that have been observed at Cordell Bank. If we now assume that the environment exerts no selection at *any* taxonomic level, it is possible to use the species fraction alone to predict the fractions of the California population at all taxonomic levels (higher than species) that would be expected on Cordell Bank. How well do the observed fractions agree with such predictions?

The data [Abbott and Hollenberg, 1976; Silva and Moe, 1978; Silva, 1981] are shown in the table on page 75. As an example from the table, 52 genera were found on Cordell Bank from the total California population of 283 genera (fraction 52/283=0.18).

Using the observed fraction of species (0.08) and the known distributions of taxa in the California population [Abbott and Hollenberg, 1976], Kerstein calculated the expected fractions of classes, orders, families, and genera, and compared them to the fractions observed on Cordell Bank (table).

ALGAE OBSERVED ON CORDELL BANK
(Compared with all algae known from California)

	No. Observed		Fraction	
Taxon	California	Cordell Bank	Observed	Expected
Classes	5	4	0.80	0.90
Orders	23	13	0.57	0.52
Families	65	18	0.28	0.48
Genera	283	52	0.18	0.18
Species	669	57	0.08	-

We find, with one exception, that the observed fractions for each taxonomic level are consistent with purely random selection from the California population. They are consistent with the hypothesis of no selection of that level in the Cordell Bank populations. Note, however, that this agreement cannot be taken as evidence of the absence of selection, since there are many possible forms of selection that could accidentally generate the same set of fractions.

The single exception is at the family level, where the observed fraction of algal families is significantly less than one would expect statistically. It appears that the conditions on the Bank are favorable to some families but not others. Certain families will find the environment hospitable and will contribute many species. Others will find it inhospitable and will contribute few or none.

Young specimens of *Nereocystis leutkeana.* These plants were not collected intact at Cordell Bank; they grew in culture from material that was collected there. The larger blade is 12 cm long.

A typical alga found at Cordell Bank, *Platythamnion heteromorphum* (magnification 40x). In relatively deep water, the red algae thrive better than the brown and green algae. This is due to the evolutionary adaptation to the color of the available light—in deep water there is only blue and green light, which is efficiently absorbed by the red pigments. Several species and genera of algae previously unknown were found at Cordell Bank, providing support for the idea that isolation does indeed influence the populations—to some degree it truly is an underwater island.

Scales of calcium carbonate from the yellow-brown alga *Coccolithus* (magnification 1700x). These interesting structures are produced by several families of yellow-brown algae grouped together under the name coccolithophorids. Because of their durable skeletal remains, they were preserved as fossils and provide a valuable stratigraphic record. The scales, or coccoliths, usually form a globular, almost spherical, case that falls apart upon the death of the cell.

A Small Place

The data in the table on page 75 also illustrate a general aspect of a limited habitat: As an organism becomes more specifically defined, the likelihood of finding it in the habitat decreases. The Cordell Bank data bear this out: The fraction of California classes, orders, families, genera, and species represented on the Bank decrease monotonically. This trend is exactly what one would expect of an island population in equilibrium with a larger parent population [Carlquist, 1974]. The more specialized an organism is, the narrower is the range of conditions, and therefore localities, in which it can live. In a small and limited locality such as Cordell Bank, we find a relatively small subset of the available species, namely those that are particularly adapted for living there. Again, these data do not provide evidence that the California population is the parent of the Cordell Bank population; but they are consistent with such parentage.

Even if the conditions are hospitable for a species to live on the Bank, it is by no means certain that it would actually do so. The very limited area provides room for a very limited number of dominant species. Furthermore, we would expect the populations to fluctuate in time: two species may be equally suited for residence, but only one (or none) is currently present. This "twinkling" of populations is somewhat akin to musical chairs, or the constant shifting done by 100 birds on 85 roosts. All these processes would reduce the number of species present on the Bank at any given moment, but we cannot easily separate the different effects.

Other Places

Another comparison that can be made is between Cordell Bank and other rocky banks. There is a bank 5 km off Pt. Sur, south of Monterey, that is almost identical to Cordell Bank in its depth and topography. The general appearance and biota of this bank are very similar to Cordell Bank. Both banks are well offshore, relatively free of coastal runoff. Both banks have an area within the 50 m contour about 5 square km. Both banks have a relatively flat top, punctuated with extremely rugged, jagged ridges and pinnacles reaching to about 40 m minimum depth. Both banks have large colonies of *Allopora*, and many sponges, mollusks, and echinoderms, and support large populations of rockfish.

In spite of these similarities, the two banks turn out to have quite different communities. Of 186 species known so far at the Pt. Sur Bank [Schmieder, 1989], 108 (55%) were known from Cordell Bank. Within the 10 dominant taxa, the overlap was 50%. This small commonality was at first surprising; what could cause such differences in the populations when the physical environments are so similar? The answer is probably the same as discussed earlier in reference to the algae: the species/area relationship. Tiny islands simply cannot support a large number of species because statistical fluctuations cause a relatively high extinction rate. The smaller the islands, the fewer species will be found, although there may be many other species available from the larger pool that could live there just as well. Which species are present at any particular time is merely a matter of chance, a snapshot of a twinkling community.

Farther south, there are several seamounts in southern California, including Cortez and Tanner Banks (west of San Clemente Island) and Farnsworth Bank (off Catalina Island). The communities on these banks are similar in many ways to that at Cordell Bank [Lewbel, et al., 1981]. Among the most abundant organisms are *Allopora, Corynactis, Strongylocentrotus, Loxorhynchus,* and various sponges. Among mollusks, 7 out of 16 (44%) bivalves (clams, scallops), 24 of 59 (41%) prosobranch gastropods (snails), and 5 of 14 (36%) opisthobranch gastropods (nudibranchs) observed on Cortez and Tanner Banks are also known at Cordell Bank. Thus, to the extent that mollusks are representative of the communities, there is about a 40% overlap between the biota on the rocky banks in southern California and that on Cordell Bank.

On the other side of the continent, Stellwagen Bank at the entrance to Boston harbor bears similarity to Cordell Bank: it is about the same size and depth, and likewise benefits from upwelling. Principal interest is in the fish, turtle, and cetacean populations, which in turn depend on the algal and invertebrate community.

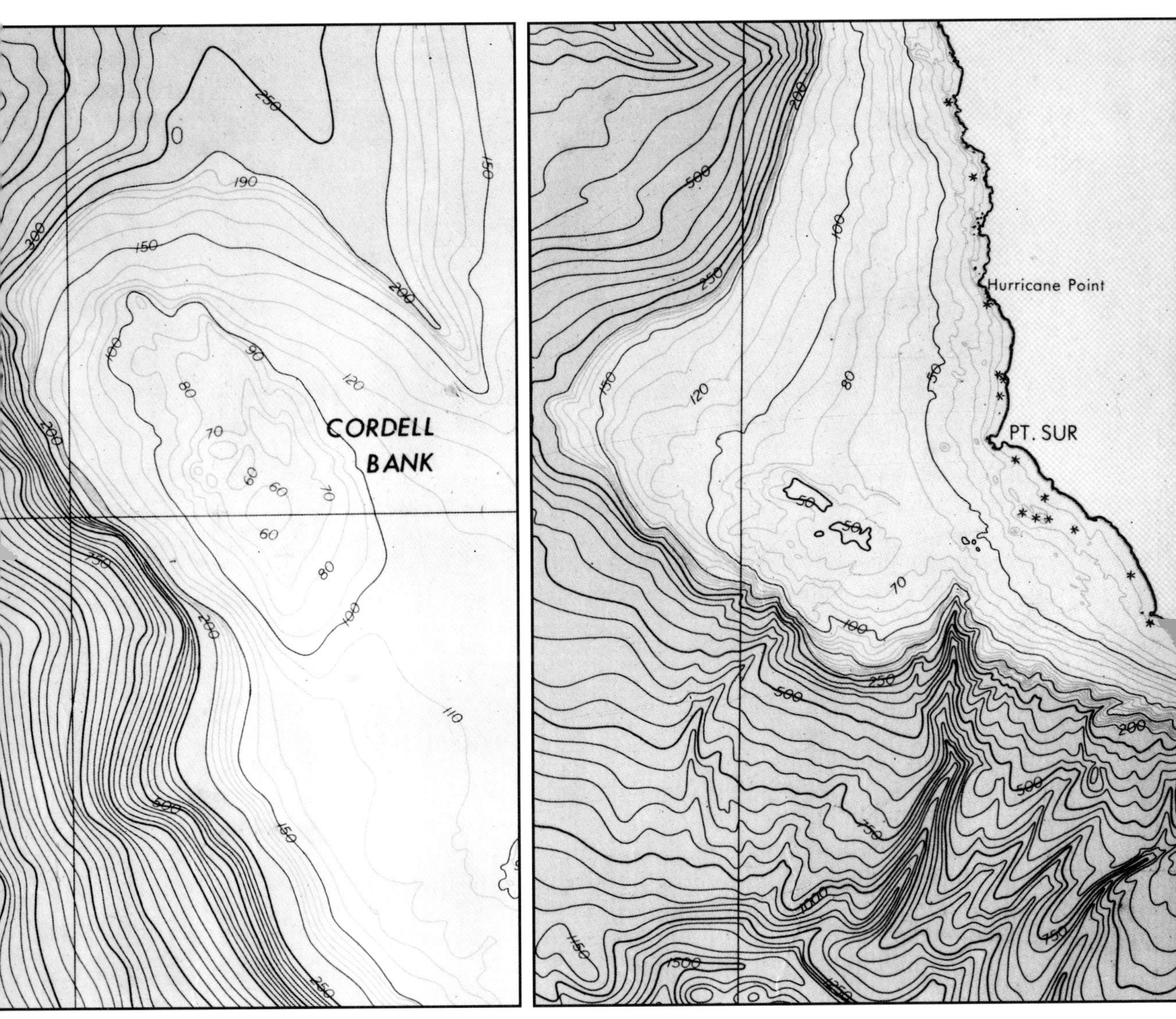

Bathymetric charts of Cordell Bank and the bank off Pt. Sur. Both banks are hard rock substrates, both have very rugged topography sculpted by surf erosion to roughly equal depths, and both lie sufficiently far offshore to have some degree of isolation. However, significant differences were observed in the populations on the two banks, almost certainly an expression of the "twinkling" of the community due to the very small area it occupies. It is likely that at another time we would find roughly the same numbers of species on the respective sites, but many of the species would be different from those we find today.

Provinces and Zones

Finally, the importance of the latitude of Cordell Bank should be noted. Various authors [Newell, 1948; Valentine, 1966; Newman, 1979] have shown conclusively that in the northeastern Pacific there are two large biogeographic provinces, called the Oregonian and Californian provinces. The provinces overlap near Pt. Conception (Lat. 35°N) where there is a marked change in the water temperature. The biota in each province is distinct. Within the transition zone between the two provinces there are more species than are found on either side of the zone. What is more important is that some species are found only within the transition zone. These so-called "short-range endemics" have been used as evidence of the narrowness of the zone, and to argue that the zone was compressed during the post-Pliocene movement of the associated land masses (less than 2 million years ago).

The relevance of Cordell Bank to the transition zone is that the northern limits of many organisms, particularly mollusks, are given as Monterey Bay, which is therefore assumed to be the northern limit of the zone. This assumption yields a relatively narrow transition zone in which the number of species is enhanced. But in the collection of 106 shelled gastropods from Cordell Bank, 32 are northward range extensions, most of which are north from Monterey Bay. There are only 2 southern range extensions among the gastropods. Another organism, the barnacle *Armatobalanus nefrens*, was until recently unknown north of Monterey. It was used as a bellwether species in defining the transition zone. These facts suggest that Monterey has been artificially emphasized as a northern limit, probably because it's so much fun to go there to collect specimens! In reality, the Gulf of the Farallones, with Cordell Bank its northward limit, may represent a more natural boundary for many species. If so, the transition zone should be considered somewhat wider and less pronounced than it currently is pictured. More data will be needed before this question can be clarified.

Range extensions for some organisms observed on Cordell Bank. For various locations along the coast, lines indicate the maximum range formerly known for various species. Arrows show how the boundaries were moved when the species was observed on Cordell Bank. For instance, one particular species was previously known only as far south as Forrester Island, but now is known to be distributed as far south as Cordell Bank. Similarly, four other species were known only as far north as Los Angeles, but now are known as far north as Cordell Bank.

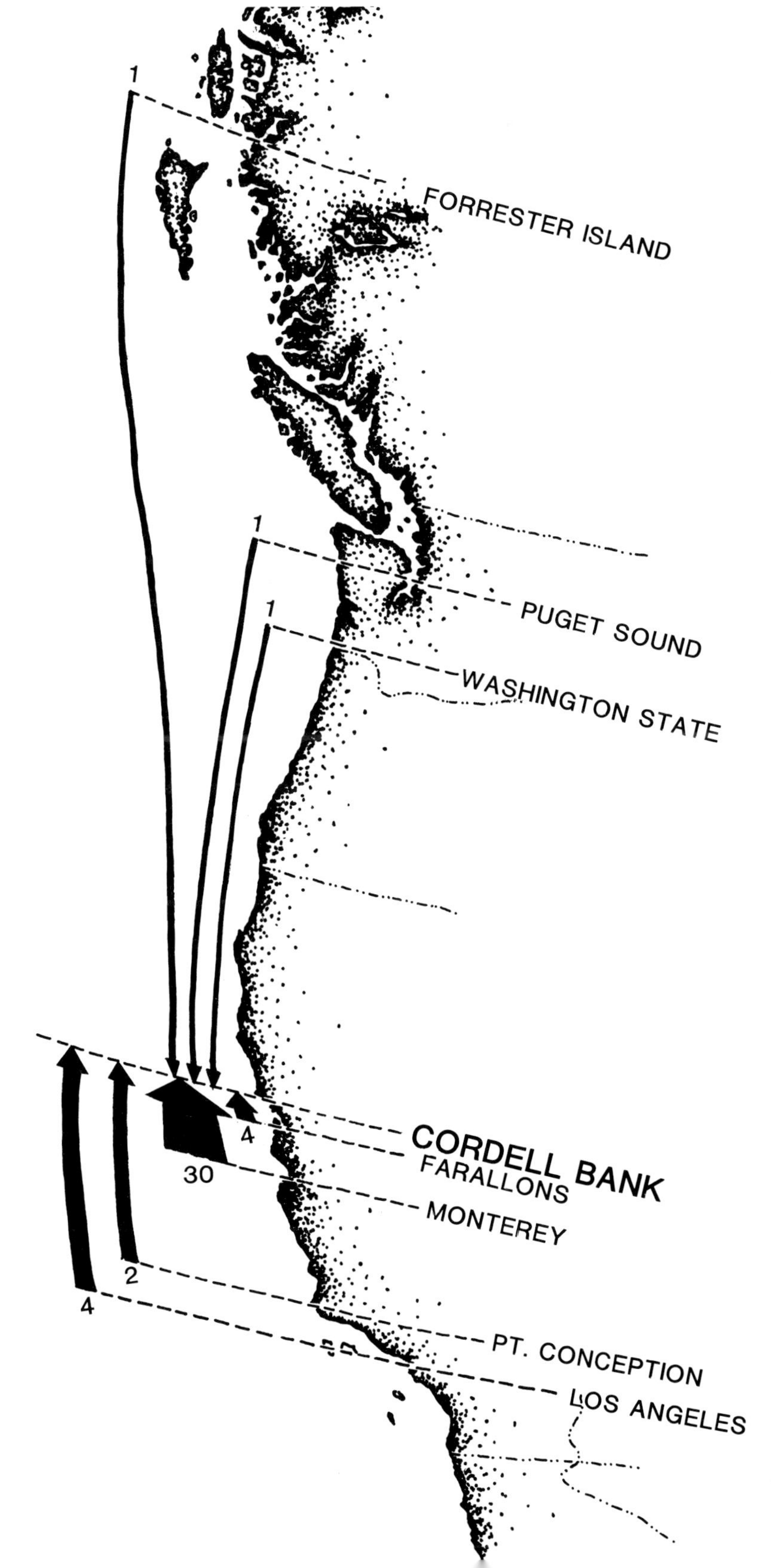

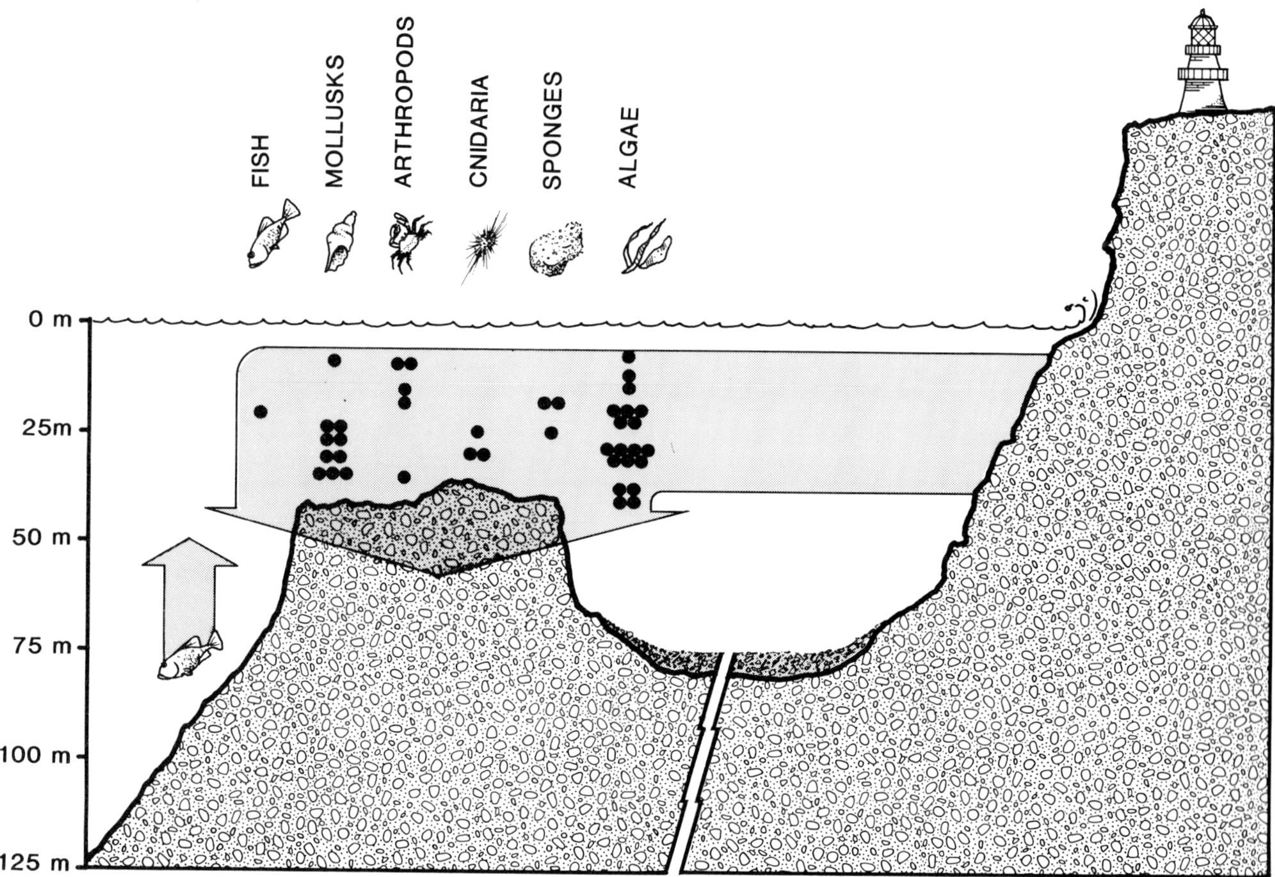

Depth extensions of some organisms observed on Cordell Bank. Each dot represents one species, grouped within major phyla. The location of the dot indicates the maximum depth to which that species was formerly known. The large arrow indicates that from the occurrence on Cordell Bank, these species are now known to live deeper, typically about 40-50 m. The small arrow indicates that one fish was formerly known only from deeper depths, and thus constitutes a depth extension to shallower water.

THE FUTURE

Artifacts

What will happen to the biological community at Cordell Bank? The answer, of course, is "It depends ...."

It may well depend on the amount and kind of human pressure that is brought to bear on this fragile ecosystem. As we shall see, there are several potential dangers to the biota. If we successfully navigate between these dangers, the evolution of the ecosystem will be naturally determined, and a few very general predictions can be made.

There is, at present, very little evidence of human activity on Cordell Bank. Divers report seeing practically no evidence of the intense fishing that is done there. One might have expected to see all manner of lead weights, line, nets, and perhaps entire rigs formerly owned by frustrated fishermen! Instead, there is almost nothing visible but the showering sponges, highrise hydrocoral, aggregating anemones, creeping crabs, and ascending algae! The reason for this pristine appearance is almost certainly that the rate of growth of the cover far exceeds the rate of deposition of artifacts. It's as if the organisms wait eagerly for the next piece of junk to arrive, whereupon they leap on it, claim it, cover it, and remove it from sight.

There are some rather startling artifacts that are hard to miss: a number of very large holes cut directly into the bedrock. Most of the holes lie deeper than 45 m, and measure about 1 m in diameter and up to 3 m deep. Most of them are almost perfect right circular cylinders. Naturally, there has been endless speculation concerning their origin. No one has come forward to claim credit for constructing them. There are many natural mechanisms by which holes can be drilled in rocks, but none seems acceptable as reasonable in the environment of Cordell Bank. At present, it appears most likely that they were made in the late 1960's by the U. S. Navy using shaped explosive charges, as part of a project to develop submarine listening devices. Most of the evidence of the terrible peripheral damage that must have been done by the explosive charges which presumably blasted these holes is now gone. It appears that the healing process is reasonably rapid in this vigorous environment.

The holes provide a valuable scientific opportunity: If they were indeed constructed at a known time, subsequent observations should give information about the dynamics of recovery from trauma—colonial succession, rates, and so on.

Damage

One exception to this picture is the hydrocoral *Allopora californica*. These colonies are extremely brittle and fragile, grow at the glacial rate of about 1 cm per year, and stand in the most exposed uppermost portions of the ridges and pinnacles. One can easily imagine a heavy object dragging over the tips of these ridges, or up and down their steep sides, shearing off every colony larger than a few cm high, taking with it all the organisms that live on the *Allopora*, including the sponges, tunicates, worms, snails, and barnacles. Following such a disaster it would take perhaps 25 years to reestablish the *Allopora* and associated community to its normal level [Gotshall and Laurent, 1979].

Fishing

Mechanical damage is not the only danger to Cordell Bank. A large fishing industry thrives in the area around and on the Bank, supporting both commercial and recreational fishermen. On the whole, this industry is balanced and self-regulated and exerts quite reasonable pressure on the fish populations on the Bank [Heimann and Carlisle, 1970]. But in the early 1980's, the pressure increased quite suddenly. By the end of 1984, so many adult rockfish were being taken from Cordell Bank that the old-time fishermen predicted significant reduction of the fish stocks within one or two years. True to the prediction, during the author's expeditions in 1986, far fewer rockfish were seen. The defense against this kind of disaster will have to lie in the increased collection of data on populations and conditions, and the vigorous execution of the statutes governing the management of our marine resources.

Oil

Another potential danger that is much discussed is oil exploration and drilling. Although there is no oil on Cordell Bank, there may be some in the Bodega Basin, just a few km upstream. At various times, tracts in the Basin have been mapped for future lease sales, and in 1985 one engineering firm was granted a permit to drill a series of exploratory wells, much to the consternation of Bay Area conservationists [Pimsleur, 1984]. The dangers to the biological community at Cordell Bank from oil activities are related to influx of sediments and toxic chemicals [Cole and Clark, 1982; Spies, 1983; Teal and Howarth, 1984]. The large populations of seabirds and marine mammals on the Bank would be especially susceptible to floating hydrocarbons.

The establishment of the Cordell Bank National Marine Sanctuary [Federal Register, 1989] is a step toward recognition of the unique resources and value of the Bank, and the need for their protection. But even though oil and gas activities are completely banned within the sanctuary, danger still exists: the 1989 Alaskan oil spill focused attention on the fact that regardless of where the spill occurs, currents can carry oil great distances. A large spill at Pt. Arena, for instance, would inevitably put oil on Cordell Bank. Thus, it is not enough to set aside protected areas; we must also protect these areas from *incoming* dangers.

Living Through It

But what if none of these acute dangers comes to pass, or the effects of such activities are relatively benign? What will happen in the long run, say in the next million years? First, we are still emerging from the last ice age; sea level is still rising (by most estimates a few millimeters per year). Cordell Bank will be covered with more and more water, presenting even more stress to the plants and animals that even now find themselves in water too deep for comfort. One by one, the last representatives of various species will lose their grip and be carried away, leaving a new mark in the "extinct" column. Only the organisms that really like deep water, such as the red algae, certain sponges, the brittle stars, the meiofauna, a few arthropods, and the rockfish, will stick it out.

But if the planets continue to turn as they have for eons, the winds eventually will change, and cooler times will return. The thirsty glaciers will again drink deeply from the oceans, and the waterline will again slip down. The rocks on Cordell Bank will again sense the sky, and light-loving plants and animals will slowly return. Thousands of years from now, Cordell Bank may look much like it did thousands of years ago. Long after humankind has become a vanished species, the granite spires will again break the surface, and Cordell Bank will once again be a real island!

(Overleaf) The ocean is replete with underwater islands. They exist on the tops of seamounts, banks, guyots, and even flat places that are subject to some differentiation from the rest of the ocean. Being an island is the rule; everywhere in the world is, in some sense, isolated from the rest of the world. At the same time, of course, everywhere in the world is also connected to everywhere else. One task of oceanic research is to determine the extent of this isolation and connectedness.

KURIL TR.
HOKKAIDO
EMPEROR
CHINOOK F.Z.
TROUGH
SURVEYOR F.
NORTH
PACIFIC
NORTHWEST
PACIFIC
BASIN
MENDOCINO
SHATSKY
RISE
SEAMOUNTS
HESS RISE
MUSICIANS
SEAMOUNTS
MURRAY
OCEAN
MID-PACIFIC
MOUNTAINS
MOLOK
HAWAIIAN
ISLANDS
TRENCH
MARIANA
BASIN
MAGELLAN
RISE
CENTRAL
PACIFIC
BASIN
LYRA
BA.
ONTONG-JAVA
PLATEAU
NAURU
BASIN
ELLICE
BASIN
PENRHYN
BASIN
MANIHIKI
PLATEAU
CORAL
SEA
FIJI
PLATEAU
SAMOA
BASIN
N. CALEDONIA
BASIN
S. FIJI
BASIN
NORFOLK
LORD HOWE
RISE
TONGA TRENCH
SOUTHWEST PACIFIC
BASIN
SOUTH
CHALLENGER
PLATEAU
TASMAN
ABYSSAL
PLAIN

ZONE
DELGADA FAN
PIONEER F.Z.
MONTEREY FAN
ZONE
BAJA CALIF.
SEAMOUNT
ZONE
FRACTURE
PROVINCE
GULF of MEXICO
FRACTURE
ZONE
ZONE
FRACTURE
CLIPPERTON
GUATEMALA BASIN
COCOS RIDGE
MARLENE
GALAPAGOS PLATEAU
BAUER BASIN
GALAPAGOS RISE
PERU-CHILE
TUAMOTU F.Z.
RISE
PERU BASIN
MENDANA F.Z.
NASCA RIDGE
QUIROS F.Z.
EASTER
FRACTURE
ZONE
PACIFIC
CHILE F.Z.
CHALLENGER F.Z.
PACIFIC
CHILE
EAN
AGASSIZ F.Z.

EPILOGUE

"Gimme a sounding," he snapped. The lead socked home in thirty fathoms ... He stopped writing ... Something told him he was right where he wanted to be. Instinctively he knew ... they were on the Cordell Bank.

—*Voyage*, by Sterling Hayden, 1976

BIBLIOGRAPHY

Abbott, I. A. and G. J. Hollenberg. 1976. *Marine Algae of California.* Stanford University Press, Stanford.

Anderson, D. L. 1971. The San Andreas Fault. In: *Continents Adrift and Continents Aground; Readings from Scientific American.* W. H. Freeman & Co. (1976), pp. 87-102.

Barber, R.T. and F. P. Chavez. 1983. Biological consequences of El Niño. *Science* **222**:1203-1210.

Barnes, R. D. 1974. *Invertebrate Zoology.* W. B. Saunders, Philadelphia.

Calow, P. 1981. *Invertebrate Biology: A Functional Approach,* Halsted Press, New York.

Cane, M. W. 1983. Oceanographic events during El Niño. *Science* **222**:1189-95.

Carefoot, T. 1977. *Pacific Seashores.* Univ. of Wash. Press, Seattle.

Carlquist, S. 1974. *Island Biology.* Columbia Univ. Press, New York.

Carlson, P. R. and D. S. McCulloch. 1974. Aerial observations of suspended sediment plumes in San Francisco Bay and the adjacent Pacific Ocean. *J. Res. U. S. Geol. Survey* **2**:519-526.

Carlson, P. R. and D. R. Harden. 1974. ERTS observations of surface currents along the Pacific coast of the United States. In: P. R. Carlson, et. al., *Principal Sources and Dispersal Patterns of Suspended Particulate Matter in Nearshore Surface Waters of the Northeast Pacific Ocean.* ERTS Final Report, 1 Sep. 1972 - 1 Jan. 1974, Publ. Feb. 10, 1975. Nat. Tech. Inform. Serv. E75-10266. 145 pp.

Carson, R. 1958. *The Sea Around Us.* Golden Press.

Chase, T. E., B. A. Seekins, and J. D. Young. 1981. Physiographic Diagram of the World Ocean's Seafloor. U. S. Geol. Survey.

Chesterman, C. W. 1952. Descriptive petrography of rocks dredged off the coast of central California. *Proc. Calif. Acad. Sci.* **27**(9):359-374.

Cole, H. A. and R. B. Clark (Eds.). 1982. The long-term effects of oil pollution in marine populations, communities, and ecosystems. *Phil. Trans. Roy. Soc. London, B. Biol. Sci.,* **297**:183-443.

Conomos, T. J., D. H. Peterson, P. T. Carlson, and D. S. McCulloch. 1970. Movement of seabed drifters in the San Francisco Bay estuary and the adjacent Pacific ocean: A Preliminary Report. *U. S. Geol. Surv. Circ.* 637B. 8 pp.

Cox, C. B. and P. D. Moore. 1980. *Biogeography: An Ecological and Evolutionary Approach.* J. Wiley & Sons, New York.

Davis, A. S., D. A. Clague, W. B. Friesen, and J. R. Hein. 1989. Composition of basaltic glasses dredged from seven seamounts offshore southern California on R/V *Farnella* Cruise F7-87-SC. Preprint.

Dewey, J. F. 1972. Plate Tectonics. In: *Continents Adrift and Continents Aground; Readings from Scientific American.* W. H. Freeman & Co. (1976), pp. 34-45.

Dueker, C. W. 1970. *Medical Aspects of Sport Diving.* A. S. Barnes & Co, Cranbury, N. J. 232 pp.

Dvorak, D. D. 1983. Private communication (underwater photographs).

Emery, K. O. 1960. *The Sea Off Southern California.* J. Wiley & Sons, New York.

Federal Register. 1981. Placement of Cordell Bank, Calif., on the marine sanctuary list of recommended areas. *Federal Register,* **46**(168):43731.

Federal Register. 1983. Announcement of Cordell Bank as an active candidate for national marine sanctuary designation. *Federal Register,* **48**(127):30178.

Federal Register. 1987. Cordell Bank National Marine Sanctuary Regulations. *Federal Register,* **52**(167):32563-32568; Environmental Impact Statements; Availability. EIS No. 870294, Draft NOAA, PAC, CA, Cordell Bank National Marine Sanctuary, Designation and Management Plan, Pacific Continental Shelf. *Federal Register,* **52**(167):32601; Public Hearings on the Draft Environmental Impact Statement/Plan for the Proposed Cordell Bank National Marine Sanctuary. *Federal Register,* **52**(177):34698.

Federal Register. 1988. Findings Regarding the Issuance of a Notice of Designation for the Proposed Cordell Bank National Marine Sanctuary. *Federal Register,* **53**(251):53049-53050.

Federal Register. 1989. Solicitation of Comments on National Marine Sanctuary Permit Application. *Federal Register,* **54**(38):8373; Cordell Bank National Marine Sanctuary. Notice of National Marine Sanctuary designation; final rule; and summary of final management plan. *Federal Register,* **54**(99):22417-22425; Availability of Final Environmental Impact Statement. *Federal Register,* **54**(51):11274; Cordell Bank National Marine Sanctuary. Notice of Proposed Rule [extending oil/gas ban to entire sanctuary]. *Federal Register,* **54**(99):22449-22451; Cordell Bank National Marine Sanctuary. Notice of Rulemaking. *Federal Register,* **54**(244):52342-52343.

Federal Register. 1990. Cordell Bank National Marine Sanctuary Regulations. *Federal Register,* **55**(233):49994.

Feldman, G., D. Clark, and D. Halpern. 1984. Satellite color observations of the phytoplankton distribution in the eastern equatorial Pacific during the 1982-83 El Nino. *Science* **226**:1069-1071.

FRG. 1978. Reconnaissance survey of the Farallon Islands area of special biological significance. Report of The Farallons Research Group, Oceanic Society, Ft. Mason, San Francisco.

Galloway, A. J. 1977. Geology of the Pt. Reyes Peninsula, Marin County. *California. Calif. Div. Mines and Geol. Bull.* 202.

Garretson, F. 1978. Visit to a Long Lost Island. Oak. Trib., Aug. 20.

Gotshall, D. W. and L. L. Laurent. 1979. *Pacific Coast Subtidal Marine Invertebrates: A Fishwatcher's Guide.* Sea Challengers, Los Osos, CA.

Hanna, G (not "G."!) D. 1952. Geology of the continental slope of central California. *Proc. Calif. Acad. Sci.* **27**(9):325-358.

Hayden, S. 1976. *Voyage.* Avon Books, New York.

Heezen, B. C. and C. Hollister. 1964. Deep-sea current evidence from abyssal sediments. Lamont Geological Observatory Contrib. 700. Reprinted in: *Marine Geology,* **1**:141-174.

Heimann, R. F. G. and J. G. Carlisle. 1970. The California Marine Fish Catch for 1968 and Historical Review 1916-68. State of California, The Resources Agency, Department of Fish and Game, Sacramento. 70 pp.

Howard, A. D. 1979. *Geologic History of Middle California.* Univ. of Calif. Press, Berkeley.

Jennings, F. D. and R. A. Schwartzlose. 1960. Measurements of the California Current in March 1958. *Deep-sea Res.* **7**:42-47.

Johnson, J. W. 1953. Sea and Swells. in: *Inshore Survey, San Francisco Bay: Literature Survey,* Univ. Calif. Dept. Eng. Rept. 57, Issue 1, p. P-30.

Kerstein, A. R. 1985. Private communication (statistical analysis).

Kruse, W. 1984. Private communication (underwater photographs).

Kruse, W. A. and Schmieder, R. W. 1986. High Resolution Graphic Images of EEZ Data: Cordell Bank, California. Proc. Fourth Working Symposium on Oceanographic Data Systems, IEEE Computer Soc., Scripps Inst. Oceanography, La Jolla, CA. 4-6 Feb. 1986. pp. 4-13.

Lewbel, G. S., A. Wolfson, T. Gerrodette, W. H. Lippincott, J. L. Wilson, and M. M. Littler. 1981. Shallow-water Benthic Invertebrates on California's Outer Continental Shelf. *Mar. Ecol. Prog. Ser.* **4**:159-168.

Limbaugh, C. and W. J. North. Fluorescent Benthic Pacific Coast Coelenterates. *Nature* **78**:497-498.

Littler, M. M., D. S. Littler, S. M. Blair, and J. N. Norris. 1985. Deepest known plant life discovered on an uncharted seamount. *Science* **227**:57-59.

MacArthur, R. H. and E. O. Wilson. 1967. *The Theory of Island Biogeography.* Princeton Univ. Press, Princeton.

Mattinson, J. M. 1982. Private communication (rock description and dating).

McLean, J. H. 1985. Private communication (mollusk identification).

Menard, H. W., Jr. 1960. Possible pre-Pleistocene deep-sea fans off central California. *Geol. Soc. Amer. Bull.* **71**(8):1271-1278. Reprinted in: Bailey, E. H., Ed. 1966. Geology of Northern California. *Calif. Div. Mines and Geol. Bull.* 190.

Milliman J. D. and K. O. Emery. 1968. Sea levels during the past 35,000 years. *Science* **162**:1121-23.

Mooers, C. N. K. and A. R. Robinson. 1984. Turbulent jets and eddies in the California Current and inferred cross-shore transports. *Science* **223**:51-53.

Moore, H. B. 1958. *Marine Ecology.* J. Wiley & Sons, New York.

Morris, R. H., D. P. Abbott, and E. C. Haderlie. 1980. *Intertidal Invertebrates of California.* Stanford Univ. Press, Stanford.

Newell, N. D. 1948. Marine molluscan provinces of western North America: A critique and a new analysis. *Proc. Amer. Philos. Soc.* **92**:155-166

Newman, W. A. 1979. Californian Transition Zone: Significance of Short-Range Endemics. In: *Historical Biogeography, Plate Tectonics, and the Changing Environment,* J. Gray and A. J. Boucot, Eds., Oregon State Univ. Press, pp. 399-416.

Nicol, J. A. C. 1960. *The Biology of Marine Animals.* Interscience Publishers, New York.

NOAA. 1982. Navigational Chart No. 18645 (formerly C&GS Chart No. 5072). San Francisco to Pt. Arena.

NOAA/NOS. 1974. Bathymetric Map No. 1307N-11B. Vicinity Pt. Sur to Pt. Reyes.

North, W. J. 1976. *Underwater California.* Univ. of Calif. Press, Los Angeles.

Owen, R. W. 1980. Eddies of the California Current System: physical and ecological characteristics. In: *The California Islands: Proceedings of an Interdisciplinary Symposium,* D. M. Power, Ed., Santa Barbara Museum Nat. Hist., Santa Barbara, pp. 237-263.

Page, B. M. 1982. Migration of the Salinian composite block, California, and disappearance of fragments. *Amer. J. Sci.* **282**:1694-1734.

Pimsleur, J. L. 1984. Marin ecologists upset over drilling plan. S. F. Chronicle, Aug. 27.

Powell, D. C. 1964. Fluorescence in the sea anemone *Anthopleura artemisia. Bull. Amer. Littor. Soc.* **2**:17.

Rasmussen, E. M. and J. M. Wallace. 1983. Meteorological aspects of the El Niño/southern oscillation. *Science* **222**:1195-1202.

Reid, J. L., Jr. 1959. Oceanography of the northeastern Pacific Ocean during the last ten years. *Calif. Coop. Oceanic Fisheries Investig. Repts.* **7**:77-90.

Reid, J. L., Jr. 1962. Measurements of the California Countercurrent at a depth of 250 meters. *J. Mar. Res.* **20**:134-137.

Reid, J. L., Jr. and R. A. Schwartzlose. 1962. Direct measurement of the Davidson Current off central California. *J. Geophys. Res.* **67**:2491-97.

Ricketts, E. F., J. Calvin, and J. W. Hedgpeth. 1985. *Between Pacific Tides.* Rev. by D. W. Phillips. Stanford Univ. Press, Stanford.

Ross, D. C. 1978. The Salinian Block--A Mesozoic granitic orphan in the California Coast Ranges. In: D. G. Howell and K. A. McDougall, Eds., *Mesozoic Paleogeography of the Western United States.* SEPM, Pacific Section, Pacific Coast Paleogeography Symposium **2**:509-522.

Roth, B. 1978. Private communication (mollusk identifications).

Schmieder, R. W. 1978. Records of the Cordell Bank Expeditions are contained in a series of unpublished reports by the author. Copies are held in the library of the California Academy of Sciences, Golden Gate Park, San Francisco, CA. 94720.

Schmieder, R. W. 1980. Intermediate forms and range extension of *Pedicularia californica* and *Pedicularia ovuliformis. Veliger* **22**:382-384.

Schmieder, R. W. 1982. A preliminary summary of knowledge of Cordell Bank, California. Unpublished MS, Feb. 24, 53 pp.

Schmieder, R. W. 1984. Cordell Bank: Marine Sanctuary Candidate Rises from Obscurity. *Oceans* **17**(4):22-25.

Schmieder, R. W. 1985. Cordell Bank Expeditions 1978-79. *Nat. Geog. Soc. Res. Repts.* **20**:603-611.

Schmieder, R. W. 1987. Terraces, Tilting, and Topography of Cordell Bank, California. *California Geology* **40**(11):258-264.

Schmieder, R. W. 1988. Cordell Bank: An Oceanic Marvel. *Defenders* **63**(3):24-29.

Schmieder, R. W. 1989. The 1988 Expeditions to Pt. Sur: Summary of Results. Cordell Expeditions Report CE-89-4, June 1.

Silva, P. C. and R. L. Moe. 1978. Private communication (algae identifications).

Silva, P. C. 1981. Private communication (algae identifications).

Simberloff, D. 1983. When is an island community in equilibrium? *Science* **220**:1275-1277.

Sources. Records of recent history are preserved in the National Archives (Washington, D. C.), the Bancroft Library (University of California, Berkeley), and the California Academy of Sciences (San Francisco).

Spies, R. 1983. The biological effects of petroleum hydrocarbons in the sea: Assessments from the field and microcosms. Rept. prepared for Federal Interagency Committee on Ocean Pollution, Development, and Monitoring. Lawrence Livermore Laboratory, Livermore, June, 1983, 108 pp.

Suess, E. and J. Thiede, (Eds.). 1983. *Coastal Upwelling: Its Sediment Record.* Plenum, New York.

Surveys. 1873: Hydrographic Chart H-1298a (Resurvey of the Bank by Ferdinand Westdahl); 1911: H-3224 (Part of a larger survey offshore from San Francisco to Pt. Arena); 1929: H-4993 (The first survey using acoustic depth sounders); 1960-62: H-8566 (A small part of a larger survey); 1985: Part of EEZ survey (A complete high resolution survey using multi-beam swath system).

Taylor, J. D. 1981. The evolution of predators in the late Cretaceous and their ecological significance. In: *The Evolving Biosphere,* P. L. Forey, Ed., Cambridge Univ. Press, Cambridge, England.

Teal, J. M. and R. H. Howarth. 1984. Oil spill studies: A review of ecological effects. *Environ. Management* **8**:27-44.

Tyler, J. E. and R. C. Smith. 1970. *Measurements of Spectral Irradiance Underwater.* Gordon and Breach, New York.

Valentine, J. W. 1966. Numerical analysis of marine molluscan ranges of the extratropical northeastern Pacific shelf. *Limnol. and Oceanog.* **11**:198-211.

Valentine, J. W. 1973. *Evolutionary Paleoecology of the Marine Biosphere.* Prentice-Hall, Englewood Cliffs.

Webber, M. A. and S. M. Cooper. 1983. Autumn sightings of marine mammals and birds near Cordell Bank, California, 1981-82. Rept. publ. by Cordell Bank Expeditions, 4295 Walnut Blvd., Walnut Creek, CA 94596.

West, R. G. 1969. *Pleistocene Geology and Biology.* Longmans, Green, & Co., London.

Zahl, P. A. 1963. Fluorescent Gems from Davy Jones' Locker. *Nat. Geog.* **124**:260-271.

ILLUSTRATION CREDITS

Where more than one person made significant contributions to an illustration, the page number is entered following each person's name.

Tom Chase	86-87 [from: Chase, et al., 1981]
Alice' Davis	70 [from: Davis, et al., 1989]
Don Dvorak	58,59, *Plates 15,27-31,33-39,42-44,49-52,55-59,61,66,69,73-75*
Sue Estey	*Plates 22,32,40,47,60,67,70,86*
Keith Flood	*Plate 48*
Terri Klinger	75
Bill Kruse	7,9,21,23, *Plates 2-12,17-18,20-21,23-24,53,62-64,72,76-80,87-94,97-98*
Rene Martin	*Plate 13* (lower) [from: Carson, 1958]
Mary McGann	61
H. L. Menard	14 [from: Menard, H. W., Jr., 1960]
Steve Metzger	*Plates 41,54*
Tom Millington	*Plates 95,96*
Richard Moe	10,76,77,96, *Plate 26*
Rob Morris	*Plate 25*
NOAA	4-5,79 [from: NOAA/NOS, 1974], *Plate 1*
Frank Pinnock	29 [Oakland Tribune, Aug. 22, 1978]
Doug Platz	19
Charles Powell	13,62 (below)
Lyn Raible	66-69
John Santilena	*Plate 65*
Randy Schmieder	17,26,28,32-33,38,40,41 [after Mooers and Robinson, 1984], 42 [after Johnson, 1953],46,55,65 (lower), 66-69,73,81-82
Robert Schmieder	3,12,17,18-19,20,24,26,28,36,37,38,43,49,52,55,57,62 (above), 73,81,82, *Plates 13* (upper),*14,45-46,71,81-85*
Jerry Seawell	8, *Plates 16,19,68*
Lew Stark	48,51,56,65 (upper)

The montage facing page 1 was assembled from articles appearing in the following newspapers: *Oakland Tribune, San Francisco Chronicle, San Rafael Independent Journal, Contra Costa Times, Valley Times, Tri-Valley Herald,* and *Sacramento Bee.*

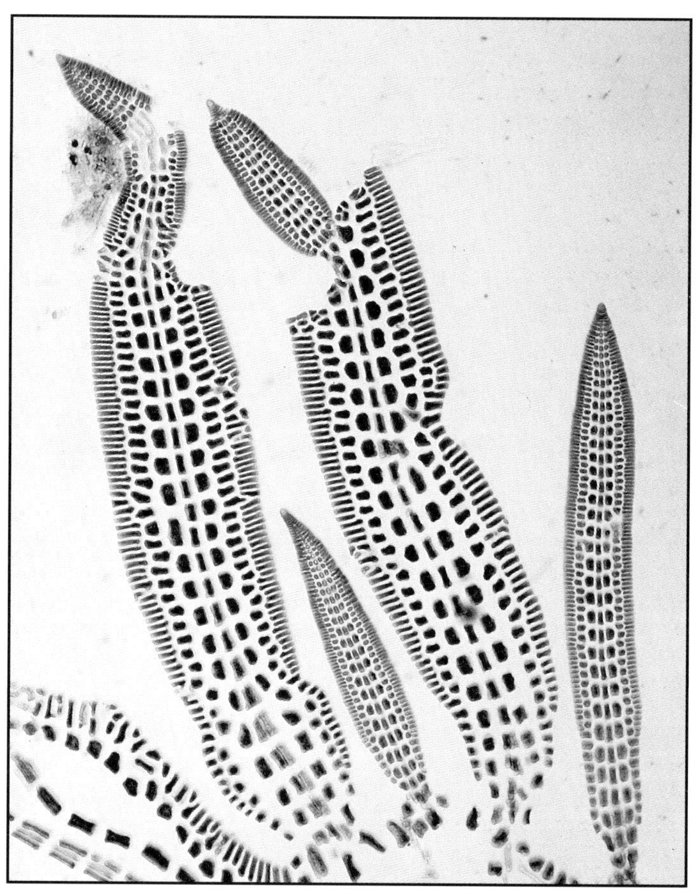

A photomicrograph of the red alga *Platysiphonia decumbens,* its cellular structure emphasized by staining to increase the contrast. The longest blades here are but 0.1 mm long. The subtlety of the cell structure and complexity of this tiny and relatively simple plant are illustrative of the staggering complexity of a community such as that residing on Cordell Bank. The closer we would look, the more we would see, to the point that we would soon run out of names, paper, and time. Any relatively complete description of such a place would occupy perhaps a million books the size of this one. This richness tells us that we have much to learn and much to care for.

INDEX

100% of the profits from the sale of this book go toward support of the research program of Cordell Expeditions.

Cordell Expeditions is a tax-exempt, nonprofit research association under the laws of the U. S. Federal Government and the State of California.

A series of reports on the scientific work of Cordell Expeditions is available. Contributions are tax-deductible. Inquiries about future research work are invited.

Please contact the publisher:

CORDELL EXPEDITIONS
4295 Walnut Blvd.
Walnut Creek, CA 94596
USA